The God of Nature

THEOLOGY AND THE SCIENCES

Kevin J. Sharpe, Series Editor

TITLES IN THE SERIES

Nature, Reality, and the Sacred
Langdon Gilkey

The Human Factor
Philip Hefner

On the Moral Nature of the Universe
Nancey Murphy and George F. R. Ellis

Theology for a Scientific Age
Arthur Peacocke

The Faith of a Physicist
John Polkinghorne

The Travail of Nature
H. Paul Santmire

God, Creation, and Contemporary Physics
Mark William Worthing

Unprecedented Choices
Audrey R. Chapman

Whatever Happened to the Soul?
Warren S. Brown, Nancey Murphy, and H. Newton Malony, editors

The Mystical Mind: Probing the Biology of Religious Experience
Eugene d'Aquili and Andrew B. Newberg

Nature Reborn
H. Paul Santmire

Wrestling with the Divine
Christopher C. Knight

Doing without Adam and Eve
Patricia A. Williams

Nature, Human Nature, and God
Ian G. Barbour

In Our Image
Noreen L. Herzfeld

Minding God
Gregory R. Peterson

Light from the East
Alexei V. Nesteruk

Participating in God
Samuel M. Powell

Adam, Eve, and the Genome
Susan Brooks Thistlethwaite, editor

Bridging Science and Religion
Ted Peters and Gaymon Bennett, editors

Minding the Soul
James B. Ashbrook

The Music of Creation
Arthur Peacocke and Ann Pederson

Creation and Double Chaos
Sjoerd L. Bonting

The Living Spirit of the Crone
Sally Palmer Thomason

The God of Nature

Incarnation and Contemporary Science

Christopher C. Knight

FORTRESS PRESS
MINNEAPOLIS

THE GOD OF NATURE
Incarnation and Contemporary Science

Library of Congress Cataloging-in-Publication Data
Knight, Christopher C., 1952–
The God of nature : incarnation and contemporary science / Christopher C. Knight.
p. cm.
ISBN: 978-0-8006-6221-9 (alk. paper)
1. Providence and government of God—Christianity. 2. Miracles. 3. Incarnation. I. Title.
BT135.K58 2007
231.7—dc22
2007020817

The paper used in this publication meets the minimum requirements of American National Standard for Information Sciences—Permanence of Paper for Printed Library Materials, ANSI Z329.48-1984.

Manufactured in the U.S.A.

11 10 09 08 07 1 2 3 4 5 6 7 8 9 10

To Cathie Clarke
astrophysicist, natural philosopher, and beloved wife

Contents

Preface

In many countries, contracts with insurance companies still refer to incidents such as earthquakes and lightning strikes as "acts of God." Few of us, however, take this as anything other than a legal archaism. Not only have the acts of God referred to in insurance contracts long since been consigned to the realm of the natural, but even when we are faced with incidents that seem to elude explanation in terms of "laws of nature," we presume that such an explanation is likely, sooner or later, to be found. Almost instinctively, we manifest what has been called the scientist's "presumption of naturalism."

For those of us who are committed to the Christian faith, this immediately presents a problem. The concept of an active and loving God, of the kind presented in our scriptures and tradition, is not easy to reconcile with the presumption that whatever happens is due to natural causes. This tension affects not only the way in which we express our beliefs, but even our basic spirituality. Our life of prayer, for example, is inevitably affected by what we think we are doing when we intercede for others or petition for ourselves. If we expect God to interfere with the natural world in order to bring about what we ask for in these prayers, we are doing one thing. If we see this aspect of our prayer in some other way, we are clearly doing something rather different. Thus, the question of how God acts in the world is not an abstract one, of the sort that has no impact on the ordinary believer. What we believe the answer to be will affect each of us in a fundamental way.

One kind of answer to this question is based on the strictest form of the presumption of naturalism, in which all causality is held to be due to the regularities of the natural world. This presumption about causality is not intrinsically atheistic, according to some who adhere to this position, since there is, they argue, nothing illogical about believing in a God who has chosen to act in the world solely through upholding the "fixed instructions" or "laws of nature" that God has given to it. A thoroughgoing theistic naturalism is, they insist, perfectly coherent.

For most Christians, however, this kind of approach provides only a part of the truth. Certainly, they admit, it is proper to see the world and the laws it obeys as important aspects of God's loving care. Nevertheless, they go on, the "general providence" that arises from the way in which God has created the world must be supplemented—at least from time to time—by some other mode of action in which God acts more directly. Unless we invoke such "special providence," they assert, we are unable to affirm the ongoing experiences of the Christian community, and much of our worship and spirituality becomes meaningless. A strong theistic naturalism may not, they admit, be incoherent. All the same, they argue, it must be seen as providing an inadequate picture of the relationship between God and the world. In the face of some event that cannot be explained straightforwardly, the presumption of naturalism may be useful as a counterbalance to the tendency to resort uncritically to explanation in terms of God's "special" action. However, they insist, this presumption must be seen as a pragmatic rule of thumb rather than as an absolute prescriptive rule.

The more perspicacious of those who take this conservative position may be aware of some of its problems. They may, for instance, recognize that it is difficult to understand why special providence does not occur more often than it apparently does. They may also recognize that a world requiring acts of special providence is worryingly reminiscent of the sort of badly designed machine that needs to be tinkered with occasionally in order to keep it functioning properly. Despite these worries, however, defenders of special providence feel they have made the right choice. The problem of divine action, they insist, hinges on the question of how God can respond to events in the world in the way that the ongoing experiences of the Christian community seem to require. Even if they put a great deal of emphasis on naturalistic processes—and in this sense might qualify as "weak" theistic naturalists—these people argue that it is theologically inadequate to focus exclusively on these processes in the way that "strong" theistic naturalists do. At the very least, they insist, there must be some kind of "causal joint"—a point of interaction between God and the world—that allows God to manipulate natural processes in such a way that a practical divine response to the world can be put into effect.

Many of these conservatives also go on to claim that there are further, and more specifically theological, objections to a strong theistic naturalism. Christians, they insist, must believe in a "special" mode of divine action, not only to account for the total experience of divine providence throughout the ages, but also to proclaim the reality of the loving, involved God who was made flesh in Christ. In particular, they suggest, this "making flesh"—as articulated in the doctrine of the incarnation—represents the supreme example of how God, who is beyond nature, interferes in a special way with natural processes.

In this book, I examine the views of this conservative majority and take, as my starting point, my broad agreement with them. The strong version of theistic naturalism has, I believe, hitherto been deficient in precisely the ways they suggest. However, I shall go on to argue, the conclusions that these conservatives draw from these insights are based on shakier foundations than they usually recognize. In particular, I shall stress two points.

The first is that there is, at a philosophical level, a fundamental flaw in the widespread belief that a strong theistic naturalism entails a very limited scope for divine providence. It is possible to construct a version of this position that is very different from that which is best known: the deism of the eighteenth century. A contemporary version of a strong theistic naturalism can be constructed in such a way that all that has usually been attributed to special providence—including events usually deemed miraculous—can be understood within a unified model.

The second point I shall make is that this perspective is strongly reinforced when the doctrine of the incarnation is understood in its fullness. The view of the incarnation that is usually adopted by those who think of themselves as conservative Christians is, I shall suggest, only a very attenuated form of the far richer notions of incarnation that have a claim to be truly traditional. These traditional notions enable us to construct a strong theistic naturalism that is hardly recognizable as such: one in which the laws of nature that can be investigated through the scientific method represent only a "low-level" aspect of the way in which God's presence in the world allows God's will to be accomplished.

This approach will seem strange to many, because the doctrine of the incarnation has (especially in Western Christianity) been diluted to the point at which its implications for our understanding of the cosmos have been effectively obliterated. The Pauline conception of the cosmic Christ has been eclipsed, as has the full meaning of the Fourth Gospel's proclamation of the divine Word, through whom all things are created. As a result of this eclipse, our reaction to the natural sciences has usually failed to see how the scientists' naturalistic perception of the universe might be interpreted christologically. Once this interpretation is made, our Christian faith may be seen in a perspective that will, for many, be entirely new. Not only does an incarnational view of the divine presence in all created things hold the key to a solution to the problem of God's action; it also provides a way of responding positively to the other faiths of the world, and of developing a spirituality that is consonant with the deepest wellsprings of our human nature.

Those who are familiar with the academic literature will recognize that some of what I will say takes its bearings from arguments presented in my previous book, *Wrestling with the Divine*,[1] and in articles I have written for various journals and conference proceedings.[2] I do not, however, assume that the reader has any knowledge of these earlier studies, nor do I simply expound their arguments in a simplified form that is appropriate to the wider audience at which the present book is aimed. Rather, my earlier thoughts have been recast in terms of my growing awareness of the Christian tradition of incarnational thinking, especially of the development of this thinking that occurred in the Eastern part of the Christian world during the early centuries of our faith.

Some might see the centrality of this approach as meaning this book can be categorized as an example of the theology of the Eastern Orthodox Church. Although I have been a member of that church for the last few years, many of my fellow members will have doubts about whether this is an appropriate description. For most of them, the term *Orthodox theology* connotes something very limited in scope: essentially an exposition of the thinking of the church "fathers" of the early centuries, in which the expositor's creative role is limited to adding a few footnotes here and there.

My own view, however, is that no matter how high this patristic thinking stands in our esteem, our present age needs far more than a few footnotes, especially when it comes to articulating a coherent response to the questions thrown up by the natural sciences. In the face of these questions, members of various Western churches have, over the last generation or two, led the way for all of us. These pioneers may not always have had a profound awareness of patristic insights. They have, nevertheless, had the courage to tackle these questions with a patristic boldness, and in the task they have set themselves, they have done much of the preliminary spadework.

To acknowledge this is not, however, necessarily to accept the sort of consensus that has emerged among these pioneers about the future of their project. As the reader will see, I myself am critical of major aspects of their approach, despite my profound appreciation of all they have attempted. An important aspect of this critique is my sense of the continuing relevance of Eastern patristic perspectives, which, in my judgment, provide answers to many of the questions these pioneers have raised. We need, I believe, not merely a revision of existing views on the interaction of science and theology, but also a new synthesis in which these patristic perspectives are given their full weight.

I cannot, of course, claim to have articulated this synthesis fully in what follows, and in some respects can claim to have done no more than lay some of the foundations that will be necessary. Much further work will be needed, and this will require that all of us, whether we are of the East or of the West, be willing to learn—as someone has put it—"to breathe with both lungs." Given the history of division and suspicion between the Eastern and Western Christian traditions, this lesson will not be easy. It promises, however, to be rewarding. For the outcome of giving both traditions their due weight will be, if I am right in my conclusions, nothing less than the development of a spirituality and theology that are truly appropriate to the scientific and pluralistic age in which we live.

1

Contemporary Science and the Religious Response

Some fifteen billion years ago, according to the current scientific account, our universe began its existence in the "big bang." From the clouds of gas that resulted from that event and its immediate aftermath, the first stars eventually formed through gravitational contraction, and in their cores, new chemical elements were synthesized through nuclear fusion. In the death throes of these first stars, these elements were spread into the interstellar medium. New generations of stars were born from this enriched medium, and they too, as they died, added to the store of the heavier elements that were needed if complex chemical compounds—and ultimately we ourselves—were to be possible.

Some five billion years ago, the star that we know as the sun came into being. Like other stars in the early stages of their formation, it had around it a disk of gas, in which were embedded particles of dust and of ice formed from the core materials of stars that had already died. And as with other newly forming stars, some of this disk material around the newborn sun eventually coalesced into planets. One of these planets was our Earth, whose mass, orbital distance, and other attributes were such that it developed conditions in which ever more complex molecules could form and eventually become self-replicating. Life had begun, and among the creatures that developed from this first speck of living material, beings with intelligent self-consciousness eventually emerged.

The most successful of these creatures were intelligent enough, and survived for long enough, to allow a complex culture to develop, in which the whole of the development from the big bang onward came to be understood with remarkable scope and precision. For the physical scientists among these humans, the character of the inanimate universe came to be seen as a straightforward result of the laws of physics acting on the initial conditions that existed after the big bang. For the biologists among them, the species of their planet, including the one to which they themselves belonged, came to be seen as a straightforward result of processes just as naturalistic as those described by the physical scientists.

The intellectual adventure that brought about these overlapping descriptions had begun slowly. In ancient and medieval times, the Chinese, the Greeks, and the Arabs had all made important discoveries. It was only in modern Europe, however, that this process accelerated to a remarkable extent, especially through two key figures who lived almost two centuries apart in England. The first was Isaac Newton, who in the seventeenth century explained planetary motion in terms of laws of motion and a simple gravitational law. What Newton had begun would, in our own age, become an astonishingly robust and beautiful explanatory framework for the entire nonliving cosmos. The second key figure was Charles Darwin, who in the nineteenth century explained the evolution of species in terms of the interaction of chance mutations with environment. What Darwin had begun would, in our own age, become an astonishingly robust and beautiful framework for understanding the entire living world.

Together, these two frameworks—those of physics and biology, together with their related subdisciplines—provide a naturalistic paradigm for understanding what we call God's creation. We cannot, of course, regard the picture they currently present as an absolutely definitive one, for the intellectual adventure goes on. There are still loose ends to be tied up, and we cannot entirely rule out a future scientific revolution, in the sense in which Thomas Kuhn has popularized the term.[1] The universe envisaged by Newton has already been incorporated, through such a revolution, into the wider vision of Albert Einstein, and the neo-Darwinism of Darwin's successors one day might in a comparable way be incorporated into a wider vision. Nevertheless, we live in a world in which the vast majority of scientists—religious believers as well as others—are confident that future developments of scientific understanding will confirm, rather than overturn, the basis of our current scientific understanding, at least to the extent that a fundamentally naturalistic picture will remain.

In the face of this broad scientific consensus, religious believers have sometimes found themselves in a quandary. Some, admittedly, have recognized that nothing in the scientists' description requires a denial that we live in a universe designed and sustained in being by God. Many, however, have been less than happy with the scientific picture of the universe's development through the interaction of physical law and chance processes. And relatively few of them have, like Arthur Peacocke, been able to affirm wholeheartedly that "God creates the world *through* what we call 'chance' operating within the created order, each stage of which constitutes the launching pad for the next."[2]

The reasons for this discomfort are often complex, but one aspect of the problem arises from the way in which Western culture has been profoundly influenced by an intellectual tradition that developed in a sophisticated way as early as the twelfth century: that of "natural theology." In this tradition, it was held that the existence of God could be indicated—or even proved—without reference to God's self-revelation in historical acts or in individual religious experience. In particular, the form of this natural theology that developed from the mid-seventeenth century onward held that the universe itself, as perceived by the scientist, could be seen as indicating the existence of God. The interrelatedness and intricacy of the world were seen as pointing unambiguously to its creation by a benevolent intelligence. As William Paley explained it in his book *Natural Theology* (1802), just as the mechanism of a watch you found would convince you of its design and manufacture by an intelligent watchmaker, so the character of the universe should convince you of the existence of its creator.

So successful was this form of natural theology, especially in the English-speaking world, that religious belief came to be seen by many as relying on its validity. The reality of God became, for many, less a matter of inward experience or of historical revelation than an intellectual hypothesis—one whose validity, it was thought, could be proved or at least rendered highly probable. Therefore, when the widespread acceptance of Darwin's evolutionary insights caused such arguments to be relegated to the waste heap of intellectual history, a major crisis occurred. Although some Christians welcomed these insights openly and joyfully, many accepted them either slowly and reluctantly or not at all. Now that the interrelatedness and adaptation of the parts of the world could be seen as the outcome of the interaction of environment and chance mutation, it was widely felt that a major plank had been removed from the platform on which the theistic hypothesis had stood.

In the wake of this, theologians understandably came to put less reliance on arguments from natural theology than previously, and many even came to view the whole tradition of such argument with suspicion. The spiritual dangers of seeing theistic belief primarily as an intellectual hypothesis were more clearly recognized, and—especially in light of the specifically theological arguments formulated by Karl Barth and others—many came to believe that the tradition of natural theology was fundamentally flawed in conception. A new emphasis on God's self-revelation in history became characteristic of much (especially Protestant) theology.

However, while the more sophisticated believer came to see the tradition of natural theology as no more than a questionable and historically conditioned form of apologetic, Western culture remained profoundly influenced by it. The less sophisticated believer, together with the nonbeliever, still tended to see natural theology's arguments from design as a necessary component of theistic belief. The undermining of these arguments' scientific basis was therefore seen by many not as the end of a dubious intellectual fashion, but as an undermining of theism itself. Because many people have equated naturalism with atheism, assumptions persist among many atheists, right up to the present day, that disproof of theistic belief requires no more than a demonstration that naturalism is valid, while some religious believers see defense of their faith as requiring an attack on naturalistic understandings of the basic structure and character of the universe.

The result of this situation has been a somewhat comic (or tragicomic) battle, not unlike that of Tweedledum and Tweedledee. On the theistic side, there has been a frantic attempt by some to update Paley's kind of argument by insisting that only "intelligent design" can account for certain components of the universe. (Eyes and wings have been the components most commonly cited.) On the other side, popular accounts of the neo-Darwinian scheme are often consciously antireligious, so that Richard Dawkins's books *The Blind Watchmaker* and *Climbing Mount Improbable*,[3] for example, expound the scientific case against the concept of "intelligent design" in what is essentially an atheistic campaigning spirit.

If we were to judge the matter entirely in terms of this somewhat unreal battle, then we would simply have to acknowledge the triumph of atheism. At the purely scientific level, Dawkins's books constitute a lucid and persuasive popularization of the overwhelming case for the fundamental tenets of neo-Darwinism. His arguments indicate that, at the very least, the neo-Darwinian scheme can be regarded as a theoretical framework that is robust, in the sense that it is now so overwhelmingly supported by evidence of various kinds that only extremely strong evidence—of a sort that has yet to be found—could conceivably constitute a legitimate challenge to its status in current biological thinking. By contrast, the counterarguments of opponents of this view fail to do anything more than indicate that some puzzles remain to be solved within the neo-Darwinian paradigm.

This is not to say that these critics fail entirely to raise legitimate questions. They are quite right, for instance, in suggesting that Dawkins, when he moves beyond strictly scientific issues, often manifests an inability to recognize the philosophical presuppositions that control his thinking. Even at this level, however, the legitimate points made by defenders of intelligent design are often masked either by the scientific illiteracy of the arguments in which they are embedded or else by a failure to integrate such arguments coherently into the wider argument about the theological implications of a naturalistic perspective. In particular, there seems to be little recognition among Dawkins's more vociferous critics that it may simply be his reductionism, rather than anything intrinsic to his scientific perspective, that leads ultimately to his atheism.[4]

This problem is, moreover, exacerbated by the way in which Dawkins's critics often indulge in philosophical nonsense when they draw false conclusions from the current state of evolutionary debate among scientists. Arguments between Dawkins and Steven Jay Gould, for example, are often cited by defenders of intelligent design in order to imply that the neo-Darwinian synthesis is questionable. Such reasoning simply ignores the realities of scientific practice, however: vigorous debate within a broad consensus is of the essence of scientific development. Disagreement at the level of what Imre Lakatos calls "auxiliary hypotheses" is an intrinsic part of the development and exploration of the "core theories" from which those hypotheses arise.[5] Current arguments about biological evolution in fact illustrate precisely this characteristic of scientific development, and as Gould himself has noted, it is simply inane to see his arguments with Dawkins as in some sense questioning what they hold in common.[6]

It is hardly surprising, given this background, that the argument that only "intelligent design" can account for certain components of the universe is considered by many to be an extremely poor one. This particular concept of intelligent design is, in fact, unlikely to carry weight for anyone who does not have a prior religious commitment to its validity. And this commitment, as we have seen, may have less to do with theistic belief, as such, than with a particular—and flawed—strand of theological thinking.

There does exist, however, a rather different argument about intelligent design that is far more subtle and complex, and which, as a result, cannot be dismissed so straightforwardly. This is based not on any challenge to the current scientific consensus, but on the recognition that the universe, as perceived by those who uphold this consensus, has been an extremely fruitful one. Our universe has given rise, among other things, to intelligent beings like ourselves. This very fact, some have suggested, may be seen as pointing to the intelligent design of a universe that, once in being,[7] can "make itself" naturalistically in a most remarkable way.

For example, the physicist Paul Davies, while acknowledging that the traditional forms of natural theology have proved impotent, has suggested that a new form of it can be developed precisely on the basis of a naturalistic understanding of the universe's increasing complexity. In particular, he notes, when we try to work out what the universe would have been like if its laws and fundamental constants had been only very slightly different, we are forced to the conclusion that it could never have evolved into a universe in which we, or any comparable beings, could have come into existence. A fruitful universe like our own seems, in fact, to depend on a number of extremely "finely tuned" factors. Should we not, he and others suggest, posit an intelligent creator who has carried out this fine tuning?

At first glance, this interpretation of a naturalistic understanding of the universe's development may seem extremely attractive to theists. There are, however, a number of problems, both philosophical and scientific, with using this "anthropic cosmological principle"[8] to develop a new form of natural theology. It is not even clear, for example, that the specific question on which it focuses—that of why the cosmos is so fruitful—is a coherent one. It involves, some argue, both wisdom after the event and a smuggling in of a sense of our own significance, which is, strictly speaking, irrelevant. The only proper response to anthropic arguments is, in this view, simply to note that if the universe had been one of the many unfruitful ones that seem to have been logically possible, then we wouldn't be here to ask questions about fruitfulness.

There are, moreover, scientific arguments that ours may be only one of a very large number of universes, and that what at present seem incredible coincidences might eventually be understandable scientifically. The only current candidates for a "many universes" understanding and for scientific explanation are, admittedly, both highly speculative and philosophically dubious. Many argue, nevertheless, that we should be wary of attributing too much to what might turn out to be only current ignorance.

As if considerations of this sort were not enough to make us wary of attempting to revive a form of natural theology based on the anthropic cosmological principle, we must also take into account the fact that there is now a widespread and strong suspicion of natural theology on purely theological grounds. While some theologians have simply been frightened off the concept of natural theology by its bad track record, many have been influenced by the sort of argument—put in its strongest form by Karl Barth and his followers—that any sort of natural theology is invalid (or, at any rate, must not be carried out separately from a theology based on historical revelation).[9] For this reason, the somewhat weak revival of natural theology that has occurred over the last generation or two has, as John Polkinghorne has noted, been "not so much at the hands of the theologians . . . but at the hands of the scientists."[10] It is no accident that the strongest claims for anthropic arguments come from those, like Paul Davies, who have little theological background, while theologians with no scientific background still tend to fulminate, as does Nicholas Lash, about "the fatuous illusion that we could discover or come across God as a fact about the world."[11]

A number of considerations suggest, then, that we should be wary of attempting to construct a new form of natural theology on the basis of our inability to uncover any reason that is naturalistically understandable for the fine-tuning of the universe. Many, as a result, judge as too sanguine the position of someone like Polkinghorne himself, who advocates what he calls a "revived and revised natural theology," in which anthropic considerations are seen as giving rise to a "meta-question . . . to which theism provides a persuasive (but not logically coercive) answer."[12] It would be better, in the view of these critics, simply to affirm that a naturalistic understanding of the universe's development can be taken as at least as much an indication of God's reality as a challenge to it. The perception that ours is a universe that has been able to "make itself" so fruitfully is, in this view, not so much persuasive of, as simply consonant with, its purposeful creation.

Indeed, even this position might seem overly sympathetic to theistic belief when we take into account the version of biological development that Steven Jay Gould has emphasized, in which there is little predictability in the evolutionary process.[13] For in the light of the perception that the termination of apparently promising evolutionary pathways has been common, it seems necessary to recognize that our own evolutionary path has reached its present stage only because of favorable circumstances that could, in advance, only have been judged as unlikely to occur. For some, this insight can be expanded to suggest that the naturalistic development of intelligent life in our universe was highly improbable.

However, further consideration based on probability theory suggests that this conclusion misreads the implications of Gould's genuine insights. For it is arguable in biological terms that the particular evolutionary route that led to the development of *Homo sapiens* was only one of a large number of routes that could have led to self-conscious, intelligent beings. This would mean that, while no one of these routes may have been much more probable than any other, and any particular one of them was unlikely to reach completion, the probability of at least one of them coming to fruition may still have been high.

This insight is reinforced by the biologist's concept of "evolutionary convergence," in which it is recognized that if a given ecological niche would favor some particular adaptation, then that adaptation may occur more than once and from significantly different starting points. Not only does genetic evidence point to the close relationship of animals with widely divergent physical characteristics—suggesting the scope of possible adaptation—but there do exist, or have existed, animals in similar ecological niches that have arrived at closely similar physical features through very different evolutionary routes: marine mammals and certain kinds of fish, for example. Once we have recognized that reasoning intelligence is itself an adaptive feature, we can affirm that it was likely to arise without having to assume that it could only have done so, or was most likely to have done so, through the particular evolutionary route that led to *Homo sapiens.*

It is, of course, true that self-conscious, intelligent life has emerged on our planet via only one evolutionary route. Compared with the eye, say, which seems to have evolved forty or so times to nine fundamentally different "designs," it would seem that nature finds it, so to speak, less easy to go down the intelligence route to adaptation than down some others.[14] Balancing this consideration, however, is the fact that there seems to be no good scientific reason to deny that life might have arisen elsewhere in the universe. This means that the probability that intelligent life would eventually arise somewhere in the universe is greater than merely earthly considerations indicate—so much so, indeed, that some have considered the simultaneous existence of many intelligent life forms in different parts of the universe to be not only possible, but perhaps even probable.[15]

The chance element in evolutionary processes does, of course, suggest that things could have been other than as they are. If there was, from the beginning, a broad predictability in the processes by which the cosmos evolved to its present degree of complexity, there was clearly no exact predictability. Once we have recognized, for example, that the light-sensitive organ we call the eye has evolved on earth in nine fundamentally different ways, there seems to be no good reason to insist that when intelligent beings eventually arose on our planet, it was probable that they would have eyes of the particular sort that we ourselves have. (In arguing otherwise, Simon Conway Morris—who has done more than anyone else to explore the implications of evolutionary convergence—in my judgment overstates the argument that intelligent beings would be very likely to end up with sensory apparatus and other features like our own.[16]) If we wish to see God as having created us through chance mechanisms, we need—as theologians have always insisted—a rather broad conception of what it might mean to claim that we have been created in God's "image and likeness." The particular physical form and sensory apparatus that we have cannot be seen as necessary aspects of this image, even if some of their functions might.

In the face of considerations of this kind, the response of some Christians has been to meet them—but only halfway. Let us accept, they say, that there are good scientific reasons for seeing the development of the physical universe in the way that this naturalistic account does. However, they go on to ask, has there not been a smuggling into this description of concepts that transcend what can be explained in terms of the physical universe? It may well be, they acknowledge, that at the physical level, the interaction of chance and physical law is an entirely adequate explanation of the way in which the universe has evolved to its present complex state. However, are there not aspects of this description that must be thought of in a separate category from the physical? What, for instance, of characteristics such as life, intelligence, and self-consciousness? Can these be explained naturalistically?

This rhetorical question is one that has, in the past, caused immense confusion. On the one hand, it has evoked—especially among committed atheists—a reductionistic analysis in which it is claimed that these characteristics in fact have no ultimate reality. True reality, they claim, lies at the level of atoms and molecules, and these other characteristics are nothing but "epiphenomena" that can ultimately be explained totally in terms of the properties of those atoms and molecules. On the other hand, and partly in response to this kind of reductionism, theistic apologetic has often fallen back on a kind of vitalism. This apologetic has asserted that you can't derive the animate from the inanimate or the conscious from the unconscious. These can only be understood as "extra" factors, inexplicable in terms of the particles and processes with which scientists are concerned. This line of reasoning concludes that the only explanation is that they have been added by God supernaturally to the physical aspect of the world.

In the face of the continuing influence of these kinds of argument, it is necessary to see them in their specific historical and cultural context. In particular, it is important to recognize that theists who continue to take up arguments of a vitalistic sort are not operating in a cultural vacuum, but in the context of a prevailing dualism, which, with its origins in ancient Greek thought, has been reinforced by strands of the philosophical vitalism, which (albeit in a diffuse and only semiconscious way) remains highly influential at the popular level.

Indeed, so strong is this influence that Christians, instead of subjecting quasi-vitalistic thinking to the strong theological critique that it requires, have all too often allowed culturally rooted assumptions of a dualistic kind to eclipse the biblically rooted anthropology of early Christian thought. In particular, instead of defending the traditional theological notion that humans are a unity of body, mind, and spirit, they have often maintained a view that opposes the "physical" to the "mental" or "spiritual." As a result of this dualism, they have, for example, frequently expressed the Christian doctrine of eternal life in terms of the immortality of the soul, rather than in terms of the restoration of the unity of body, mind, and spirit.

But if the questionable nature of this kind of dualism has been recognized less often than it should have been on the basis of theological arguments, such recognition is now becoming more widespread because of philosophical perspectives. The biblical anthropology that rejects any dualistic split between the physical and the mental is strongly reinforced by the contemporary philosophical critique of both vitalism and reductionism. In fact, philosophers are increasingly seeing life and self-consciousness as representing high levels of complexity in the cosmos that, while not fully explicable in terms of the components and processes found at lower levels of complexity, do not require that something be "added" to those lower levels in order for them to exist. Instead, it is now widely held, we should see these higher levels of complexity naturalistically but nonreductively, as autonomous *emergent properties*, understandable perhaps in terms of the existence of top-down organizing principles. As the physicist Paul Davies has put it:

> There is no compelling reason why the fundamental laws of nature have to refer only to the lowest levels of entities, i.e. the fields and particles that we presume to constitute the elementary stuff from which the universe is built. There is no logical reason why new laws may not come into operation at each emergent level in nature's hierarchy of organization and complexity. . . . It is not necessary to suppose that these higher level

> organizing principles carry out their marshalling of the system's constituents by deploying mysterious new forces specially for the purpose, which would indeed be tantamount to vitalism. . . . [Instead, they] could be said to harness the existing interparticle forces, rather than supplement them, and in so doing alter the behaviour in a holistic fashion. Such organizing principles need therefore in no way contradict the underlying laws of physics as they apply to the constituent parts of the system.[17]

This philosophy of emergence,[18] allowing as it does a middle way between reductionism and vitalism, has been considerably important in the dialogue between science and theology in recent years. It has not only allowed scientific insights into the physical rooting of our mental and emotional life to be accepted by theologians, but has also, when coupled with the more general perspectives on the development of the physical universe that we have noted, allowed development of a widespread consensus in which God's action as creator can be seen in terms of the natural unfolding of the potential that God has given to the cosmos from its very beginning. As a result, it is now widely acknowledged that the scientific description of the development of the universe, from its beginning to its present complex state, need in no way be a barrier to religious faith.

This belief that the early universe's potential manifested the intentions of a divine creator radically challenges many of the arguments that natural theology has used in the past. The theology that arises from this belief has often, therefore, been termed a "theology of nature" in order to distinguish it from the older approach with which, in certain respects, it stands in marked contrast.

In particular, this theology of nature recognizes, as we have seen, that there are good reasons to deny the validity of some of the apologetic arguments of the natural theology of the past. It has abandoned entirely both a simplistic view of "intelligent design" and the vitalistic assumption that naturalistic processes need supplementing in some way to bring about the existence of the attributes that specifically characterize living beings and humanity. According to the advocates of this theology, however, this abandonment does not mean we must relinquish the concept of God as creator. Rather, it means we can envisage God as creating in a way that is based on the interplay of chance and physical law. The character of the universe is, they affirm, consonant with the reality of a creator who has "designed" it with the particular aim that there should arise within it, through naturalistic processes, beings who can come to know God as their creator and redeemer.

2

Casting Down the Idols

The new theology of nature is based, as we have seen, on a picture of the process of creation that involves a naturalistic interplay between chance and the physical law that has been designed by God with a providential aim. This picture is not, as we shall see, without its problems for many Christians. It is, however, at least consonant with the extraordinary and awe-inspiring picture that the natural sciences of our time have set before us, and in certain respects, it has proved extremely helpful to believers.

It has, for example, gone a long way toward solving one of the problems that believers have often experienced: the perception that much in the natural world seems not only aesthetically unappealing but also ethically dubious. In particular, the "cruel" reliance of many animal species on the deaths of members of other species can strike believers as difficult to understand and appreciate within a religious framework. When our religious response to the world is informed by scientific perspectives, however, the questions that occur to us about this aspect of the world are considerably modified.

In particular, if we are aware of the evolutionary and ecological framework of contemporary biology, then even the ugliest or "cruelest" creature may be seen as having its necessary part to play in the totality of the world. This may not solve all our problems with respect to this issue (and we shall return to it presently). It does, however, at least remove the sense of frustration that we can experience if we try, without this framework, to integrate this aspect of the world into our theological understanding.

Indeed, an awareness of this sort will tend to make all creatures—whatever our purely aesthetic reaction to them may be—a source of fascination and of awe, and will enhance the sense of God that we have because of the more obvious beauty of other aspects of the natural world. Because of this, the "atheistic" popularization of contemporary biological insights by people like Richard Dawkins may be seen as one of the most important contributions that has been made to the spiritual life of our time.

The advantages of the new theology of nature are by no means limited to this sort of issue, however. In fact, its most important contribution is that it challenges the dissonance that is widely perceived in our culture between naturalistic perspectives and theological ones. In this sense, it is extremely important at an apologetic level, especially for existing believers, who can be reassured that theistic beliefs are not in themselves simply incoherent.

Although the new theology of nature has proved immensely valuable to existing believers, its impact on nonbelievers has proved, by and large, to be no more than marginal. Perhaps it has led some of them to acknowledge that the reality of God is not a logical impossibility. Except for those few who have perceived some degree of apologetic force in anthropic arguments, however, they have usually concluded that the new theology of nature contains nothing that can persuade them of that reality. As they often put it, God still seems to them neither a necessary hypothesis nor one whose validity may be seen as probable.

In the face of this inability to persuade, some of the defenders of the new theology of nature have understandably wondered whether their approach might be further developed so as to play a more aggressive apologetic role, comparable to that which natural theology once played. Might the perspectives that inform their theology of nature be expanded so as to indicate not simply that the "hypothesis of God" is tenable, but that it is actually intellectually preferable to the atheistic hypothesis? In particular, some have wondered whether it might be possible to defend some sort of "inference to the best explanation" scheme, in which, while no single argument is used on its own to argue the probable reality of God, a number of such arguments can be claimed to have a cumulative persuasive force.

At first sight, this approach has its attractions for the Christian. My own view, however, is that it suffers from a twofold problem that ultimately subverts it. First, arguing in terms of inference to the best explanation gives the impression that the reality of God is, for those with religious faith, essentially a hypothesis about the world that we happen to find convincing. The true nature of faith—as a trust in God that has its roots in inner experience—is eclipsed. Second, those who argue in this way fail to understand why atheists and agnostics tend to see their arguments as invalid. By suggesting, at least implicitly, that those who remain unpersuaded by their arguments are simply being irrational, they fail to recognize that being a "rational agent"[1] is far more complex than simply being obedient to logical rules.

An important element of this complexity can be seen in the aspects of scientific judgment explored by the philosopher of science Thomas Kuhn in his analysis of why, during scientific revolutions (such as the change from Newtonian to relativistic dynamics), proponents of different theoretical frameworks can find themselves at loggerheads. According to Kuhn, the answer is that all scientific theories are embedded in wider frameworks of thought and practice, which he calls paradigms. In times of scientific revolution, he says, there are not just competing theories within a single accepted paradigm, but competing paradigms, each of which affects the interpretation of data. As a result, there is no straightforward pool of data to which some neutral set of logical rules can be applied in order to decide which side is right.[2]

The details of Kuhn's argument have, admittedly, been controversial, not least because they can be interpreted as denying the very concept of scientific rationality. Most recent philosophers of science have, however, tended to repudiate the view that Kuhn's genuine insights can be extrapolated in this way. Rather, while they recognize that aspects of his analysis require modification, they see him as correct to insist that scientific practice does not consist simply of the application of a set of logical rules to some set of theory-independent data. Logical thought, they argue, is indeed central to the scientists' practice, but scientific work also—and inevitably—involves discrimination between different options in a way that goes beyond what can be described in terms of logical rules.[3] As Michael Polanyi once put it, scientific work requires not just logic but also "tacit judgment."[4]

This analogy of competing paradigms is, I believe, helpful when we try to understand how and why religious believers and nonbelievers tend to see arguments for God's reality in different ways. In terms of this analogy, the two sides represent different paradigms, so the arguments that each side tends to think of as "neutral" are, in fact, subtly but inevitably affected by the paradigm to which it holds. This makes for a situation comparable to that which Kuhn describes in terms of "incommensurability," in order to explain why scientific revolutions are rarely smooth transitions but usually involve a real inability, on the part of one side of the argument, to see the world as the other side sees it.

In terms of this analogy, those of us who are religious believers see the world in a particular way. Insofar as we see the reality of God as an explanation of aspects of the world, it may seem to us that there are arguments that validate our belief. For those who don't see the world in the same way, however, the arguments that seem so suggestive to us can offer little more than an indication that the "God hypothesis" may not be completely incoherent. It is not that one side is cleverer or more clear-sighted than the other; rather, each side sees the evidence offered in a different way, using its "tacit judgment" differently. As in the scientific case, adoption of the other side's paradigm requires not so much perspicuity as a sort of gestalt "conversion."

We need not, of course, deny that there is at least some value in the modest ability of inference arguments to challenge the atheist's belief in the redundancy of the "God hypothesis." Those of uncertain faith and those nonbelievers who—at an unconscious level—are already moving toward faith may well find the removal of this barrier significant. But even when we acknowledge this, it is important to recognize that the origin of faith never lies primarily in being convinced by intellectual arguments. From a psychological and empirical perspective, religious faith is something that arises not primarily from intellectual conviction, but from a far more complex process with a strong unconscious component. Even in those few cases in which some kind of inference argument has been instrumental in religious conversion, that argument can usually be seen, on reflection, less as a reagent than as a catalyst for those unconscious processes. This does not mean that there is no point of contact between believers and nonbelievers. It means, simply, that this point of contact does not exist primarily at the level of intellectual argument. Rather, it exists at the level of deeper aspects of our common human experience, which can be seen for what they are only through experiential conversion.

One such aspect is the human sense of awe at the natural world, which for those of us who are believers is related to our sense that the world is the creation of God, who "saw that it was good" (Gen 1:13, 21, 31). This sense of awe is not, however, limited to believers. Indeed, it is arguable that the scientific enterprise is fueled, at least in part, by the way in which those who are directly involved in it are (whatever their religious views) gripped by what Richard Dawkins has called "a feeling of awe at the majesty of the universe and the intricate complexity of life." This feeling enables them to see science as pointing to something quite other than "an existence that is bleak, devoid of meaning, pointless."[5]

This implicit sense of something intrinsically meaningful in human existence undoubtedly constitutes a measure of common ground at the experiential level between the believer and nonbeliever. The nature of this common ground can be seen more clearly, however, when it is examined in the context of another factor that fuels the scientific enterprise: the widespread sense of the intrinsic significance and value of the search for truth. For example, in the work of Dawkins himself, this sense of the intrinsic significance of truth is evident, and it gives his arguments—even those advocating atheism—a certain charm. However, it also challenges those arguments in a way that he seems not to recognize. For, since a sense of intrinsic significance and value cannot arise from scientific insights as such, it points clearly to the essentially metaphysical and nonscientific source of his intellectual motivation.

The point here is that the content of the natural sciences can never give rise straightforwardly to statements about intrinsic value or significance. (Science can use such terms only in the very restricted sense inherent in phrases such as "survival value.") To behave as if something has *intrinsic* significance or value is at least implicitly to have made a metaphysical judgment in a way that any systematic attempt to defend the atheist position must avoid. A coherent atheism can only be a form of nihilism. However, the majority of those who describe themselves as atheists are certainly not nihilistic. A sense of the intrinsic significance and value of certain aspects of human life is one that they implicitly share with believers, even if they cannot bring themselves to adopt the sort of language that believers use to point to the origin and ground of these essential elements of an authentic human existence.

If we can recognize the embryonically religious nature of the atheistic scientist's motivation, however, it is important that we do not misunderstand its scope. The problem here is well illustrated by Dawkins's own observation that, if awe at the character of the universe constituted religious belief, then he himself would rightly be considered "a deeply religious man."[6] However, he goes on to ask, if religion is "allowed such a flabbily elastic definition, what word is left for *real* religion, religion as the ordinary person in the pew or on the prayer-mat understands it?"[7]

Here—as so often when he strays beyond scientific matters—Dawkins simultaneously highlights an important question and uses the bluntest of philosophical instruments to answer it to his own satisfaction. This is evident, for example, in his comments on a fascinating study by Ursula Goodenough, in which she explores her perception of the religious significance of the awe she feels in the face of scientific insights. As far as Dawkins can see, his own atheistic views "are identical to [her] 'religious' ones," so that "one of us is misusing the English language, and I don't think it's me."[8]

In this, however, he is wrong. Labels such as "religious" cannot, either in ordinary English usage or in a philosophical perspective, be defined straightforwardly in terms of some hypothetical set of beliefs. Such labels can in fact properly be used of a spectrum of different views, which are linked to one another by what some philosophers call family resemblances. For example, when we examine the views of the ordinary Buddhist, who does not believe in God, and of the average deist, who believes in God but upholds a naturalism that is essentially identical to Dawkins's, we are well aware of the differences between them. Neither of them is very close to Dawkins's hypothetical "ordinary person in the pew or on the prayer-mat," yet to deny to either of them the label "religious" would be to misuse the language in a far more fundamental way than Goodenough herself possibly does.

This is not to deny that there is—implicitly at least—an important point in Dawkins's comment. If we can rightly claim that he is in some sense religious, then we do need to qualify this statement by acknowledging the major differences between the religious stance that he attributes to ordinary believers and that which is implicit in his practice of the pursuit of truth. These differences mean, among other things, that he or they (or both) have beliefs that require, at the very least, development and refinement. From Dawkins's atheistic perspective, many believers tend to hold a number of beliefs that are untenable. From their perspective, his passion for truth and awe at the world, while admirable, are rendered spiritually impotent by his failure to adopt a "unified theory" (as a scientist might call it) through which these existing aspects of his life can be integrated and further developed.

Dawkins himself would presumably respond to this latter view of him by saying that the religious unified theory that is urged on him is not based on evidence of the kind that would lead him to adopt a new scientific theory. And in this, we must recognize, he is right. Faith in God is, as we have noted, not the same as belief in a hypothesis that is based solely on empirical evidence. Faith is, rather, something that is entered into in a way that reflects, on one level at least, the kind of paradigm shift that occurs during scientific revolutions. For as we have seen, the analogy with scientific paradigm shifts is illuminating insofar as the entry into faith does not necessarily involve new evidence but does involve a new perception of the world.

The analogy is inadequate, however, because when someone enters into this new religious way of seeing, it involves not just a new perception, but a personal relationship with the truth that is different from the relationship that scientists have with the theories they believe to be valid. As Keith Ward has put it:

> Whereas science is a matter of tentative hypotheses, religion is about being grasped by an overpowering ideal. Science offers predictive explanation, whereas religion pursues a goal that promises to integrate all life's endeavours. Science works by continued critical testing; religion by commitment to realize its ideal vision, by trust in the power which discloses it, shows the way to it and moves one towards it.[9]

It is not for nothing that "the way, the truth and the life" (John 14:6) is understood by Christians not just as an abstract notion but also as something manifested in a person.

This kind of contrast between science and religion can, however, be overstated and needs to be balanced by recognition of the rational element in religious reflection.[10] The real achievement of the new theology of nature, for example, has been precisely to use this rational element in a valid way, demonstrating that once faith has been entered into experientially, the new perception of the world that accompanies this entry may be seen as consonant with the evidence on which contemporary scientific understanding is based. In a wider way, too, religious doctrine may be seen as providing believers with a way of understanding more deeply the various kinds of partial knowledge of the world they share with unbelievers. In this sense, religious doctrine does constitute a unified theory, even if its adoption cannot be based on an abstract consideration of its intellectual merits alone.

From this perspective, religious outreach is something that cannot simply be equated with apologetic in the old sense of the term. For atheists and agnostics to move into religious faith, what is needed is not primarily intellectual argument about the consonance of religious doctrine with scientific perceptions, or even about how the nature of their existing values can only be understood in metaphysical terms. Persuasion requires, rather, whatever engenders in people the inner experience that allows them to see—in a more direct way than rational arguments can provide—the truths that these arguments attempt to demonstrate. If we are unable to evoke this experience (to allow the "penny to drop," as the saying has it), then this may represent a failure on our part. It is not, however, a failure of intellect, but of love. As a spiritual father of the Eastern Church once said, "No one can turn away from the fallen world and approach God if he does not see on the face or in the eyes of just one man the light of eternal life."[11]

This perspective on the nature of faith has important implications, not only for our approach to those without faith, but also for our own continuing spiritual development. In particular, faith's rooting in inner experience means that to the degree we have faith in its fullness, we will have moved away from fear that it will be undermined by knowledge derived from science or from some other domain of human experience. The definition of theology as "faith in search of understanding" will seem valid to us, and we will not be fearful of the process by which the provisional understanding we have at any particular time can be refined. This does not mean our faith is irrational, in the sense that rational argument will have no effect on us. It means, rather, that we will tend to see all such argument as a means to purify our faith. We will see our basic convictions as corresponding to those scientific "core theories" that Imre Lakatos has described, which are held by scientists with a confidence in their soundness even though competing and more tentatively held "auxiliary hypotheses" inevitably exist in relation to them.[12]

Even the arguments for atheism will, if we are spiritually mature, seem helpful to us, since we will see that they may have an important role to play in purifying the way in which we express our faith. We will understand that while part of our conception of God is truly appropriate—in the sense that it is the best possible at the present stage of our spiritual development—part may be no more than a leftover from earlier stages of that development. The importance of atheistic argument here is precisely that it can help us to see such evolutionary remnants for what they are: concepts that are idolatrous because they have become substitutes for the reality of God.

Most of us, for example, can think of conceptions of God that were appropriate to a particular stage of our childhood development but that we have now outgrown. In this sense, we can see that we would be indulging in a kind of mental idolatry if we still clung to them. We also need to recognize, however, that not all inappropriate conceptions are as easy to discard as the more obviously childish ones are. In practice, there are often other, rather more subtle, conceptions that we need to outgrow but that still have their effect on us because of our spiritual immaturity. This may particularly be the case when the effects of that immaturity are exacerbated by the corporate immaturity of our own part of the Christian community.

Possibly the best way to see what I mean would be to consider the difficulties we face when we attempt to understand the historical development of the Christian conception of God. For we have been brought up, by and large, to see the scriptures as witnessing equally validly, if not always equally clearly, to the one true God. As a result, the savage tribal god of the earliest phases of our community's development, still evident in parts of the Old Testament, sometimes coexists rather uneasily in our minds with the loving, merciful, and universal God of the New Testament.

We do not have to scratch far below the surface of the scriptures, however, to see that our community's conception of God has in practice developed considerably since those early days. Even before we take the New Testament into account, we can already recognize the way in which the tribal god of the earliest period was gradually transformed, through prophetic interpretation of historical experience and of the legends that arose from that experience, into the one God of the Judaistic faith: the creator of the cosmos and the redeemer, not only of God's chosen people but, through them, of all humankind.

This process of transformation may be seen especially clearly when the books of the Old Testament are read in a historical-critical perspective that takes into account their redactional history. From this historical analysis emerges a picture of the development of Judaism in which, at each stage, relatively new conceptions coexisted with remnants from earlier periods. Just as the evolutionary biologist can still find, in any living creature, features of the way in which that creature's ancestors adapted to an environment that has long since disappeared, so the biblical historian can find, in documents of any given period, aspects of the religious attitudes of earlier periods. In particular, the transformation from a tribal god to a God with no divine rivals turns out, in this perspective, to have been a gradual and by no means early development. The strict monotheism that we tend to read into the earliest strands of the Old Testament is in fact an anachronism.

This is important because it underlines an insight that arises independently of the question of the dating of the phases of this transition: Once the conception of a single God had been developed, it retained, in some respects, characteristics that had been appropriate only to the earlier age of polytheism. In particular, just as the gods of the earlier period had been assigned quasi-human attributes that distinguished them from one another, so also the single God of the new perception was still understood as having such attributes—albeit ones that, under prophetic influence, took on an increasingly ethical dimension. Even as these elements changed in detail, however, evolutionary remnants among them tended to remain, and this was reinforced by the high status given to documents and traditions from much earlier periods. The result was the kind of belief in a single God that still bore traces of earlier belief—for example, the concept of God's "jealousy" of other gods. It was not yet radical monotheism, but a transitional attitude, sometimes referred to as henotheism.

It is notable that many of the early Christians recognized at least one aspect of the henotheistic framework that was inherent in the Judaistic background of their faith. Faced with the problem of reconciling the loving and merciful God pictured in the New Testament with the earlier and savage god of tribal warfare, some even went as far as to disown the Old Testament altogether. However, the mainstream of Christian believers found a more subtle way of dealing with this problem. They insisted that the true meaning of difficult passages was to be found through allegorical and other kinds of "spiritual" interpretation.[13] For example, when they read an Old Testament passage in which God ordered God's enemies to be slaughtered, they interpreted it as meaning simply that evil thoughts should be suppressed.

This mode of scriptural interpretation may be problematic for many of us now. At the time, however, it had an important role. Nowadays, we tend to think in terms of a progressive revelation of the nature of God, so we see the earliest conceptions we find in the Old Testament as aspects of a now discarded, but once necessary, phase of our faith's development. The earliest Christians, however, did not—and in their cultural context, could not—have this historical conception. For them, therefore, allegorical interpretation was a necessity in a way that it is not for us. Through it, the "inspired" nature of the scriptures could be defended while their literal meaning could be dispensed with whenever that seemed appropriate. In this way, their henotheistic aspects could be dealt with in a way that aided, rather than retarded, spiritual growth. (Indeed, much of the early Christians' allegorical interpretation of scripture remains excellent spiritual reading.)

At a more fundamental level, too, the thought of the Christians of the early centuries marked an attempt to move away from the henotheistic situation they had inherited. Some of them went as far as to perceive a danger in seeing any attributes in God whatsoever. For example, the writings attributed to Dionysius the Areopagite—a strong influence on the spirituality of both East and West throughout the medieval period—were adamant about this.[14] There was a strong emphasis on the negative or "apophatic" element in anything we can say about God. Attributes that can be assigned to created things can, according to this apophatic view, be assigned to God only in a very loose, metaphorical sense. As Vladimir Lossky puts it, every concept that we use of God is in danger of being "a false likeness, an idol. The concepts which we form in accordance with the understanding and the judgement which are natural to us, basing ourselves on an intelligible representation, create idols of God instead of revealing to us God Himself."[15]

It is a sad fact, however, that these early Christian attempts to overthrow the mental idols to which we are prone never entirely succeeded. Especially in the Western part of the Christian world, they often were all but forgotten. As a result, the need to deal with these inadequate ways of thinking remains a major problem for us. We may half recognize that some of our quasi-instinctive ways of thinking about God are potentially idolatrous. At the same time, however, we are inevitably affected by the fact that many of our fellow Christians not only urge us to bow down to the mental idols to which they themselves remain attached, but also insist that, unless we do so, we do not have a true faith. As a result of this situation, we can—at a certain stage of the maturing of our faith—experience acute discomfort because we have come to suspect, in our heart of hearts, that some widely accepted set of theological propositions is in fact unworthy of belief.

This may be illustrated by a problem that most Christians have, in fact, long since solved for themselves: the question of how the Genesis accounts of the creation are to be understood. If we interpret these passages as literal, historical descriptions, there will inevitably seem to be a clash between scientific and biblical perspectives. If, however, we are happy to interpret these accounts not as literal, historical descriptions of how the cosmos came into being, but as essentially poetical or philosophical accounts of God's purposes in the cosmos, this problem simply disappears.

Here it is helpful to recognize that for Christians of the early centuries of our faith, what was primarily important in these passages was the concept of humanity being made in the image of God. The "historical" details were frequently interpreted as representing simply philosophical truths, and indeed, later evolutionary perspectives were sometimes anticipated in a remarkable way.[16] As Panayiotis Nellas has put it, what was important for this tradition of interpretation was not scientific insight as such, but the recognition that "the essence of man is not found in the matter from which he was created but in the archetype on the basis of which he was formed and towards which he tends." It is precisely for this reason, he goes on, that for a truly traditionalist understanding of the creation, "the theory of evolution does not create a problem . . . because the archetype is that which organizes, seals and gives shape to matter, and which simultaneously attracts it towards itself."[17]

This kind of profound but nonliteral interpretation is, indeed, even easier for us than it was for the Christians of the early centuries, because we have a historical perspective that was largely absent from their understanding. As well as being able to adopt or adapt patristic perspectives of the sort that Nellas expounds, we can see clearly that the Genesis accounts of creation were expressed as they were because of the prescientific culture within which they were developed. Therefore, even if we are unaware of the patristic tradition of biblical interpretation, few of us now see these accounts as contradicting the scientific account of the mechanisms through which the cosmos evolved to its present degree of complexity. Rather, we see them as pointing to God as the originator and upholder of those mechanisms. If there are still some Christian communities in which questioning a literalist interpretation of these accounts leads to an enormous sense of guilt, it is evident to us that this situation is the outcome of sociological factors inherent in the way in which these communities of biblical fundamentalists operate.

What may not be so obvious to us, however, is that our own Christian communities are inevitably influenced by sociological factors of a comparable kind. While the great majority of us do not face the particular problems that arise for biblical fundamentalists, comparable—if more subtle—situations are still likely to arise for each of us. At one or more stages of our spiritual development, we are likely to find ourselves uncomfortable because we have come to suspect the validity of some way of thinking that is widely accepted in our own part of the wider Christian community. When we reach one of these stages, we are likely to find ourselves at an impasse unless we remember that faith is not belief in some set of explicit propositions, but is, as I have indicated, a trust in God that casts out fear. The sort of questioning that necessarily accompanies this stage is, therefore, not a sign of inadequate faith. In fact, such inadequacy is more likely to be manifested as a "conservative" reluctance to face legitimate questions. Indeed, much that passes for theological conservatism is in fact rooted in fear rather than in faith.

We must be careful, however, not to assume that all theological conservatism is of this essentially neurotic kind. A maturing faith may necessarily lead toward a creative questioning of the models that have previously sustained us. It also, however, leads to recognition of the danger encapsulated in the old metaphor about the baby being thrown out with the bathwater. As we mature in faith, we will recognize more and more clearly that there is a crucial need for the discernment that can allow us to distinguish the henotheistic remnant in our beliefs from what is vital to our ongoing spiritual life. This requires something that is not primarily intellectual but more distinctly spiritual.

Here, it seems to me, some comments by the late Metropolitan Anthony (Bloom) of Sourozh are of great significance. For not only has Metropolitan Anthony been widely regarded (both within and beyond his own church) as a saintly figure of great spiritual acumen; he has also, to our great profit, made specific comments about how we should respond to the kind of doubt that comes to us because a cherished picture of God has been challenged. We should, he suggests, see this doubt as playing a creative role in our spiritual life.

Just as scientists rightly doubt the models of reality that have arisen from their work, says Metropolitan Anthony, so also there is a similar need for doubt when we are faced with challenges to our picture of God. For when a scientific model is first developed, a good scientist's reaction "will be to go round and round his model in all directions, examining and trying to find where the flaw is, what the problems are which are generated by the model he has built, by the theory he has proposed, by the hypothesis he has now offered for the consideration of others." This is possible and proper, Metropolitan Anthony goes on, precisely because the scientist recognizes the relationship between model and reality:

> At the root of the scientist's activity is the certainty that what he is doubting is the model he has invented—that is, the way he has projected his

> intellectual structures on the world around him and on the facts; the way in which his intelligence has grouped things. But what he is also absolutely certain of is that the reality which is beyond his model is in no danger if his model collapses. The reality is stable, it is there; the model is an inadequate expression of it, but the reality doesn't alter because the model shakes.[18]

This is, interestingly, an attitude with strong affinities to that which has been dominant among those who have been most influential in the development of the new theology of nature in recent years. In particular, Karl Popper's understanding of the nature of scientific development—with its picture of the way in which one scientific model succeeds another in a process that involves "increasing verisimilitude"—has strongly influenced the way in which many of them have seen Christian doctrine as involving models that, while "approximately true," are capable of being replaced in principle by new models that incorporate a nearer approximation to truth.[19]

This notion of the "approximate truth" of religious doctrine may, perhaps, disturb conservative Christians who, while not biblical literalists, still emphasize the "certainties" of the Christian revelation as though that revelation consisted simply of a series of intellectual propositions. It is important to recognize, however, that this view, like biblical fundamentalism, is not conservative in the best sense of the term. Both are relatively modern phenomena, which ignore the early Christian period's profound sense of the limitations of human language and models.

As so often, Metropolitan Anthony puts the spiritual aspect of this beautifully when he speaks about the opportunity that may be being offered to us when an old model is challenged and a new one offered. He tells us that when we say "that our intellectual, philosophical, theological, scientific model is inadequate in comparison with reality, it really means that we are saying 'How marvellous, I have come to a point when I can outgrow the limitations in which I have lived and I can move into a greater, deeper, more enthralling vision of things as they are.' " The scientist's doubt, he continues

> is systematic, it is hopeful, it is joyful, it is destructive of what he has done himself because he believes in the reality that is beyond and not in the model he has constructed. This we must learn as believers for our spiritual life both in the highest forms of theology and in the small simple concrete experiences of being a Christian. Whenever we are confronted with a crossroads, whenever we are in doubt, whenever our mind sees two alternatives, instead of saying "Oh God, make me blind, Oh God help me not to see, Oh God give me loyalty to what I know now to be untrue," we should say, "God is casting a ray of light which is a ray of reality on something I have outgrown—the smallness of my original vision. I have come to a point when I can see more and deeper, thanks be to God."[20]

It is, essentially, this "more and deeper" that the best of those who have contributed to the new theology of nature have tried to explore as part of their response to the sciences of our time. Of course, they have not always agreed on precisely what this "more and deeper" consists of, nor as I shall argue, have they usually recognized important aspects of the Christian tradition that can shed light on the task they have undertaken. Nevertheless, by pointing the way toward a theology that is both consonant with scientific understanding and, as a result, richer in its perception of God's relationship to the cosmos, the theology of nature has—for those whose faith is in God's self rather than in inadequate models of God—opened up new perspectives that have a remarkable potential.

3

The Problem of Divine Providence

Among those who have accepted the broad thrust of the new theology of nature, the question of whether they need to see God's action as creator in terms that go beyond the role as the universe's designer, initiator, and upholder has proved difficult to answer. Many, despite an informed scientific understanding, have been unwilling to see the universe's development in purely naturalistic terms.

For example, although John Polkinghorne recognizes important aspects of the truth in the naturalistic view, he is wary of some of its apparent implications. The problem for him is not, as for some others, the apparent "blindness" of chance processes, since he is well aware that such processes can—as the owners of gambling casinos know—lead to a broadly predictable outcome. He does not, however, want to deny that God has ever been involved in this development in a more "direct" way. To deny this, in his opinion, is too close to the deistic picture of an impersonal God for comfort.

At least some, however, do understand God's creative role in essentially naturalistic terms. Arthur Peacocke, for example, has spoken of God's creative role by likening it to a fugue, in which the potential of a simple musical theme is explored. For him, the recognition that the inherent creativity of the cosmos depends on "chance operating within a lawlike framework"[1] leads ineluctably to the need for theists to see God as creating "*through* what we call 'chance' operating within the created order, each stage of which constitutes the launching pad for the next."[2]

It is notable, nonetheless, that Peacocke and Polkinghorne share a great deal of common ground in their understanding of God's action in the world. Both of them see the naturalistic unfolding of the creation's potentiality as an important component of a proper view of God's action as creator. This view is set within a wider perspective, however, based on their shared assumption that there are providential actions other than those involved in the interaction of chance and physical law. For both, God's action includes a providential aspect that requires recognition of a mode of divine action that transcends the purely naturalistic.

Peacocke expresses this in terms of a simple distinction between God's creative and providential action. Polkinghorne tends to blur this distinction by seeing specific providential responses to the world within the creative process. This is merely a question of where to draw the boundary, however, since neither accepts the validity of the sort of "strong" theistic naturalism that sees divine action as limited to upholding the "fixed instructions" God has given to the cosmos. Both are conservative in the sense that they believe not only in the "general providence" that arises from the regularities of the natural world, but also in a "special providence" that arises when God "responds" to events in the world through some mechanism that, while it may make use of natural processes, is not analyzable solely in terms of those processes.

This belief constitutes the mainstream position within the new theology of nature. A naturalistic perspective is fully acknowledged as an important component of our understanding of the world yet is ultimately found wanting. God's action in the world, it is held, consists both in letting that world be itself—fulfilling its divinely given potential "on its own," so to speak—and in responding to events in that world in a way that goes beyond what a purely naturalistic perspective can describe.

One of the main insights that have been crucial to the development of this position has been precisely that which we explored in chapter 1: the widespread consensus that naturalistic processes—ones explicable in principle in terms of the laws of nature—can account adequately for the broad features of the universe in which we live. This has led to a widespread recognition that, however our understanding of divine action is developed, it must avoid the once-common approach in which this action was effectively identified with the events or processes that seem not susceptible to naturalistic explanation.

Part of this avoidance has arisen from apologetic motives. It has been recognized that what "science will never be able to explain" has in practice proved susceptible to scientific explanation time and time again. More importantly, however, it has been recognized that any approach that stresses gaps in scientific explanation has a far poorer theological justification than was recognized until relatively recently. Such an approach relies on a particular understanding of "nature" that has its roots not in anything intrinsic to Christian theology, but partly in the late medieval philosophy of the West and partly in the "scientific" development of that understanding that occurred in early modern times, through which much of the subtlety of the medieval picture was in fact lost.

The historical background to both the early development and later coarsening of this view of nature is somewhat complex and need not concern us here. What we do need to note, at this stage of our discussion, is that the rise of modern science in the seventeenth century led to a situation in which, for more than two centuries thereafter, divine action was understood largely in terms of the question of how God could be seen as interacting with the deterministic, "clockwork" universe that scientific theory seemed to posit. The "natural" world was one in which God—its designer and creator—was seen as normally being no more than an observer of its autonomous workings. As a result, the only way in which God could act directly was, it seemed, by temporarily suspending the natural laws by which the world normally operated and then intervening directly in a "supernatural" manner.

The deists of the eighteenth century were the most extreme of those who accepted this clockwork model. They saw God essentially as a sort of absentee landlord and simply denied that supernatural intervention ever took place. Divine providence, they insisted, was limited to the "general providence" inherent in the well-designed machine that constituted the universe. In the context of the science of the period, this version of theistic naturalism was understood to imply a denial of both the biblical miracles and the efficacy of intercessory prayer. Because of this denial, the majority of Christians were understandably reluctant to accept their views.

This majority was not, however, unaffected by the deistic outlook and tended to adopt an outlook that can perhaps be best described as semideism. While they acknowledged the possibility of divine intervention, they tended to assume that the "special providence" it brought about was very occasional and not to be expected on a day-to-day basis. In the eighteenth-century context, this was most commonly expressed in terms of the debate about miracles that was then at its height. While God *could* perform miraculous acts, it was often said, God in fact chose not to do so, since they had only been appropriate to the "age of miracles" that was a preparation for, and accompaniment of, God's supreme self-revelation in the person of Jesus Christ. To all intents and purposes, it was assumed God was nowadays an absentee landlord, even if God had—long ago—once been in the habit of visiting the tenants.

Deism and this semideism shared a set of presuppositions about how God could act. This was based on the "clockwork" model of the universe that was seen as an intrinsic part of Newtonian physics. Deists and semideists may have disagreed about whether supernatural intervention had ever occurred. What they agreed on was that it was the only way in which it was theoretically possible for "miraculous" events to occur. One of the results of this was that those who eschewed deism of the more extreme sort and defended the concept of "special" divine providence tended to identify divine action with events that did not seem susceptible to scientific explanation. The focus was thus on "gaps" in current scientific explanation, and God became, for all intents and purposes, what has been called a "God of the gaps."

This semideistic picture of God as a benevolent absentee landlord who occasionally visited (or at least had visited) the tenants was not entirely unquestioned. Over a century ago, for example, when many Christians objected to Darwin's understanding of evolution, some of them did so less because of a literalist interpretation of the Genesis account than because it challenged their picture of the process of creation as a series of supernatural acts.[3] As Aubrey Moore pointed out at the time, however, this picture of "special creation" was theologically questionable. In fact, he argued, the Darwinian picture can be seen as "infinitely more Christian" than the concept of a series of supernatural acts, because it implies "the immanence of God in nature and the omnipresence of His creative power." Those, he said, "who oppose the doctrine of evolution in defence of a 'continued intervention' of God, seem to have failed to notice that a theory of occasional intervention implies as its correlative a theory of ordinary absence."[4]

This "ordinary absence" of the "God of the gaps" was, however, intrinsic to the semideism that many Christians then espoused. By Moore's time, there were admittedly many varieties of this outlook, since the semideism of eighteenth-century thinking had by then often been modified by various kinds of vitalism. However, all of these forms of semideism shared a common ground: their acceptance, whether implicit or explicit, of what Moore had called God's "ordinary absence." And it is precisely against this ordinary absence that the new theology of nature has reacted, by insisting that God's immanence in the cosmos must be stressed in any valid account of divine action.

The main factor that has influenced this reaction has been the recognition that, from a scientific perspective, only a stress on God's immanent presence in naturalistic processes can allow an adequate picture of God's action as creator to be maintained. In addition, a further factor has been important. This has been the way in which physics itself—through the development of quantum mechanics in the early twentieth century—has moved away from the deterministic model of the universe implied by Newtonian physics. This development has meant that events at the submicroscopic level are now seen as inherently unpredictable, in the sense that only probabilities can be assigned to particular outcomes. This indicates that, even if the relevant laws of nature are perfectly understood, events at the macroscopic level can no longer be predicted with absolute certainty, even though some particular outcome may be said to have an extremely high probability.

For many theologians, one of the attractive features of this understanding has been that it seems to allow for "special" divine action that does not require the kind of intervention in which the laws of nature are set aside in order that God may act directly. The simplest and earliest model of how this could occur—that popularized by William Pollard[5]—was, in fact, based straightforwardly on quantum mechanical insights, with God's action being seen as occurring through the alteration of the probabilities of events at the submicroscopic level.

Since Pollard's advocacy of this model, its problematic aspects have been pointed out by John Polkinghorne and others.[6] Still, the concept of some comparable sort of temporal "causal joint" between God and the world remains a central aspect of the exploration of the interface of science and theology. If the world is not deterministic but simply probabilistic, it is held, then there can in principle be a number of ways in which God can alter the probabilities of events without setting aside the laws through which they come about.

Since this understanding involves replacing the old, supernaturalist understanding of divine intervention, it has seemed appropriate to many to "refuse the word 'intervention' . . . as the way to speak about divine acts."[7] As a result, it has become common to speak of a "noninterventionist" God, who uses the laws of nature as tools for divine purposes. The "God of the gaps" can, it is thought, cheerfully be proclaimed as dead, and the only question that remains, in this perspective, is that of the location of the causal joint that enables this type of divine action to occur.

For John Polkinghorne, for example, quantum mechanical and chaotic phenomena point to the reality of a more "subtle and supple" universe than those phenomena in themselves indicate—one in which the "cloudy unpredictabilities of physical process" can be interpreted as "the sites of ontological openness."[8] For Arthur Peacocke (who feels the need to avoid what can still be interpreted as "local" divine action), God is seen as acting on the "world as a whole" in such a way that specific, local providential events are brought about through a process of "whole-part constraint." This scheme takes its bearings from the way in which complex wholes can, in scientific perspective, be seen as having an effect on the parts of which they are made up.[9]

For me, however, neither of these causal-joint schemes is persuasive, for a number of reasons. The first is that, in both cases, an analogy rather than an actual mechanism is drawn from the physical world. In Polkinghorne's case, the use of deterministic chaos theory to posit an indeterministic aspect of the universe makes it clear that his model is less an outcome of our current scientific understanding than a theologically motivated belief that this is what the world ought to be like. Peacocke's scheme, too, is based on an analogy of this sort—albeit a more general one—and is, moreover, susceptible to a scientific critique that he has not considered. (The critique is that information, in the perspective of relativistic physics, cannot travel faster than the velocity of light. Divine response on a short time scale must therefore, in his model, depend only on what are, cosmologically speaking, the very local effects of whatever has been done to the whole cosmos.[10])

A second issue that I believe should affect our judgment of this causal-joint scheme is that an adequate model of "special" divine action requires more than an account of what is sometimes called the "basic action" of God. It also requires at least some understanding of what has been called the "instrumental substructure of God's acts"—the chain of events that leads from God's basic action to the event that is identified as God's providential aim.[11] In the causal-joint schemes that have been suggested within the dialogue of science and theology, any such understanding is conspicuously absent.

These two objections indicate, I believe, that the veneer of plausibility that the causal-joint scheme manifests should not be allowed to give the impression that the problem of divine action has been solved. It is true, perhaps, that this scheme can, as Polkinghorne has noted, envisage God as "continuously at work in a way consistent with the known laws of nature (themselves understood theologically as expressions of God's faithful and unchanging will for his creation)."[12] In this sense, it does allow a new emphasis on God's continuous upholding of natural laws in all divine acts. At other levels, however, problems remain, and the ones we have already noted are not the only ones.

Far more important than these issues is the fact that a "noninterventionist" causal-joint scheme gives the illusion of having abandoned divine interference with the world when it has done nothing of the sort. It has, in fact, simply replaced one mode of interference with the world—that in which the laws of nature are set aside—with another, in which those laws are used as tools. It is "noninterventionist" only in the weak sense that a particular *kind* of interference has been abandoned.

The fact that interference is still envisaged is absolutely clear, since those who advocate this kind of causal-joint scheme still have an implicit picture in which two outcomes of any situation are possible. One is that which nature, *left to itself* (in the sense of simply being sustained in being by God), would probably bring about. The other is that which will come about if God chooses to respond to events in the world in a *direct, temporal* way through some kind of manipulation of the laws of nature.

This picture of two essentially separate modes of divine action—sustaining and manipulating—is, we should note, characteristic even of those who, like Peacocke, see God's continuous action as creator entirely in naturalistic terms. For, as we have seen, he still speaks of "providential" action over and above what he sees as God's "creative" action, and understands this providential action in causal-joint terms. For this reason, his description of his position as being a version of theistic naturalism is questionable, since he is very far removed from the sort of "strong" theistic naturalism in which it is assumed that divine providence is to be understood solely in terms of God's upholding of the "fixed instructions" that God has built into the world. By affirming God's providential "response" to the world, Peacocke explicitly envisages bringing into play something over and above God simply sustaining the universe and its laws: a new "decision" on God's part.

This new "decision" represents, quite clearly, a new causal factor in the situation. While the *physical* causes of an event of special providence are, in Peacocke's scheme, all natural ones, at another level an additional cause is required to explain why nature, nondeterministic in character, has followed the particular path that it has. What Peacocke rightly perceives as being absent from his view of God's action as creator—a "kind of *additional* influence or factor added to the processes of the world"[13]—is explicitly present in his view of God as the author of providential events. At most, therefore, his outlook constitutes what we might call a "weak theistic naturalism."

This distinction between a strong and a weak theistic naturalism is subtle, however, and advocates of the latter are often, in my experience, far from clear about it. But even if this were not the case, their motivation for attempting to avoid a strong theistic naturalism would be understandable enough. For, as far as they are able to see, theirs is the only approach—other than that which posits supernatural intervention in the old sense—that can affirm the occurrence of events that have usually been ascribed to special providence. If they come to see their causal-joint scheme as questionable, therefore, they are unlikely to adopt a strong theistic naturalism without a struggle. Instead—like Nicholas Saunders, in the wake of his own study of some of the problems associated with such schemes—they are liable to conclude that "*contemporary theology is in a crisis.*"[14]

As we shall see, however, the kind of impasse in which such people find themselves is based on a judgment that has its roots in an inadequate conceptual scheme. For if contemporary theology is indeed in a crisis, this is not just because of the difficulty of finding a coherent model for "special" divine action to supplement what is possible through a divine sustaining of the natural world. Rather, as I shall indicate, the problem lies also in the inadequacy of the prevalent understanding of the character of the natural world, which leads to the belief that a supplement of this kind is necessary.

4

Beyond Deism

Among those who have developed the new theology of nature, the main motivation for defending the concept of special divine providence is, as we have seen, the belief that it is necessary in order to affirm the reality of events (such as divine communication and answers to intercessory prayer) that represent, in some sense, a divine "response" to what has happened in the world. One aspect of this argument has often been the insistence that God is "personal," and that personal action is intrinsically tied to the ability to respond directly to events. However, this is not the case, as is clear from the fact that even human providential action can, in principle, be planned in any one of three ways.

Consider, for example, the case of college students who are still financially dependent on their parents. The financial support of those children—an important aspect of parental providence at this stage of their lives—can in principle, and even in practice, be arranged in any one of three basic ways:

1. It can be entirely unmediated. ("Here's your regular allowance, and here's some extra cash for your car repairs.")
2. It can be entirely mediated, as through fixed instructions to a bank. ("Transfer such and such an amount every month to each of my children's accounts, to cover their general expenses, and if any of them provide documentary evidence that they have had their car repaired, transfer an extra amount to cover those costs.")
3. It can be either mediated or unmediated, depending on circumstances. ("The money that comes automatically into your bank account will only cover your everyday expenses, so here's some more for your car repairs.")

Since whatever is possible with respect to human providence is also possible with respect to the divine version, this analogy allows us to recognize that there are, at least in principle, alternatives to the standard model of divine providence, which—with its distinction between general providence and special providence—corresponds to the third of these classifications.

Once we recognize that God has alternative options, however, we must be wary of assuming that God's preferred option need be the same as that which wise humans would choose. For instance, the fact that human parents will in practice be wary of using the second option (entirely mediated providence) must be seen as irrelevant in the case of God's choice. Most humans will acknowledge that their wisdom is limited and that, therefore, no matter how carefully they have set up their "fixed instructions," the general providence that arises from them will need to be supplemented, at least occasionally, by direct action of a "special" kind. By contrast, an infinite wisdom—which is what we attribute to God—is surely quite capable of setting up "fixed instructions" in such a way that there is no need to supplement the general providence that they provide.

At least in principle, therefore, a loving God *could* have arranged divine providence entirely through the fixed-instruction mode. However, we still need to ask whether this model is anything more than a theoretically conceivable one. In particular, can anything be said about it that might make it preferable to the response model? My contention is that this is the case, and I shall argue this presently from a specifically theological perspective. Before that, however, it seems necessary to comment about the most obvious objection to the fixed-instruction model, at least in the form it has taken through using the analogy of human providence. This objection is that if God can only have set up fixed instructions as humans must—that is, in the rather clumsy form of a set of if/then statements—then we clearly have a model of divine providence that is neither elegant nor based on mechanisms that are conceivable (unless, perhaps, we invoke some kind of angelic agency). It is therefore important to recognize that this particular human analogy serves only an illustrative purpose. Precisely how God may have set up providential fixed instructions in a less clumsy way may be hard to guess. It is not altogether, however, beyond conjecture.

For example, naturalistic understandings of human psychology have often manifested characteristics that are at least suggestive of divine providence. Indeed, for some, this has been an important aspect of their understanding of the concept of providence. For Christopher Bryant, who was one of the most profound commentators on the relationship of psychology to the Christian faith, C. G. Jung's "idea of the self, the whole personality, acting as a constant influence on my conscious aims and intentions in a manner that I was powerless to prevent . . . brought home to me the inescapable reality of God's rule over my life." Bryant explains:

> So long as I thought about God's providence as an abstract truth, part of theistic belief, it made no powerful impact on me. But it was quite another matter if God's guiding hand was within my own being, within the fluctuations of mood and the ups and downs of health. . . . I came to understand that to resist God was to run counter to the law of my own being; God's judgment worked through a kind of inbuilt psychic mechanism; it was self-acting and imposed from within me.[1]

This kind of insight into one aspect of divine providence does not depend in detail on the Jungian framework that evoked it in Bryant's own thinking. Yet an interesting outcome of a specifically Jungian approach to this issue is that it provides a way of going beyond Bryant's specific point so as to incorporate providential events in the external, empirical realm. This may be done through the notion of "meaningful coincidence," which many people, including Jung himself, have been led to ponder by their own experience. For some, at least, this kind of experience has suggested that internal psychic states can have the effect of evoking or coinciding with significant external events. What Jung called synchronicity is, for such people, simply a fact of experience that they feel they must accept, and if they are naturalists, they can see no way of doing this except by also accepting Jung's conjecture that there exists "an as yet unknown substrate possessing material and at the same time psychic qualities."[2]

Those who insist on the reality of special providence will, of course, give a different explanation of meaningful coincidence when it occurs. When such coincidence appears to manifest God's response to intercessory prayer, for example, they will claim that this is precisely what it is. However, not only is a Jungian approach to such coincidence susceptible to theistic analysis in terms of general providence; it is also reinforced by an important aspect of modern physics: the connection between mind and matter that seems to be required by aspects of quantum mechanics. When David Bohm speaks of the need for the physicist to comprehend matter and consciousness "on the basis of a common ground,"[3] we inevitably hear echoes of Jung's "unknown substrate."

Whether this substrate or ground can ever be demonstrated and explored remains problematic, of course. Nevertheless, the fact that the relationship between mind and matter has arisen as a concern within both psychology and physics makes it clear that extending a naturalistic, psychological account of divine providence to empirical phenomena is far from unthinkable.

Moreover, an extension of this perspective to the sort of phenomena that are usually deemed miraculous is far from impossible. In the next chapter, we shall examine this in more detail, so at present I will simply observe that it would be straightforward to expand this kind of approach in terms of a concept that John Polkinghorne has noted in his own discussion of miracles (albeit with a different intention). As he points out, physicists are now familiar with what they call changes of "regime," which bring about new phenomena that may neither have been anticipated nor even have seemed possible. The most obvious example may be the phenomenon of superconductivity, in which, in certain materials, electrical resistivity suddenly disappears when a sample is cooled to below a certain threshold temperature. This, as Polkinghorne notes, provides a good example of how phenomena that are discontinuous with ordinary experience, and at first seem inexplicable, can be the result of underlying continuities. What we call miraculous may, he suggests, be due to an analogous change.[4]

All this is, of course, extremely speculative. However, keep in mind that my purpose in discussing these possible mechanisms for an extended general providence is not to attempt a definitive account of how all that has previously been attributed to special providence can be understood in a different way. Rather, my suggestions are intended simply to illustrate that it is possible, from within a secular framework, to conjecture mechanisms through which God might have organized general providence in such a way that this providence is not limited in the way that it is in the deistic model.

This possibility does not depend on the validity of any particular set of conjectured mechanisms but does depend on a much more general belief that is implicit in the search for such mechanisms. This belief is that the created order, with its inbuilt "fixed instructions," is far more subtle and complex than our present scientific understanding indicates. This belief might be difficult for some naturalists to accept. However, it is not incompatible with naturalism as such, nor is it a belief that theists are likely to find unacceptable in principle.

Even so, in the minds of many Christians, two theological objections to this scheme are likely. First, it will seem to them that this model must envisage God as the deists did: as no more than a distant "absentee landlord." Second, it will seem to them that this scheme cannot incorporate incarnational beliefs, since the doctrine of the incarnation, as commonly understood, posits the coming into the natural world of God, who is, by definition, beyond the natural.

The first thing to note about these objections is that they are intimately interrelated, being based on the notion—common in Western philosophical theism—that the world is intrinsically separated from God. Therefore, it is important to consider that this notion is now under widespread attack, and not only because of a reappraisal of its historical roots.[5] For a number of other reasons, too, many theologians—including a significant number of participants in the dialogue of science and theology—now advocate the position they call "panentheism," in which the world is seen as being, in some sense, "in God" rather than separated from God.[6]

One of the implications of this position is that it immediately rescues theistic naturalism from what might otherwise have been one of its main drawbacks. For it underlines most pointedly what Willem Drees has noted of a more general theistic naturalism of the "strong" kind: that God, as the ground of all reality, can be seen by the naturalist as "intimately involved in every event—though not as one factor among the natural factors."[7] A God who in some sense "contains" the world can hardly be said to be absent from it.

This point is underlined in a very particular way when we look at the second of the theological objections we have noted: that which relates to the doctrine of the incarnation. Although the notion that "the Word became flesh" is at the heart of the Christian faith, the very wide scope of this phrase is only rarely recognized. All too often, it is simply taken as an affirmation of a particular historical event: the coming into the natural world of a particular historical individual: Jesus of Nazareth. That this may be a somewhat imprecise and inadequate statement of the doctrine is, however, clear from its biblical roots. The Pauline concept of the "cosmic Christ," for example, already points toward something more subtle and universal than a simple historical understanding of the doctrine. Exegesis of the doctrine's main biblical source, the prologue of the Fourth Gospel (John 1:1–14), also suggests that only in the context of a particular understanding of the whole creation can the significance of the historical Jesus be comprehended.

The problem with this prologue is, however, that translations of it—which simply render the Greek term *Logos* as "Word"—fail to indicate to the modern reader the many nuances that the term would have had, both for the Gospel's author and for its original readers or hearers. As Andrew Louth has observed (in the wake of Theodore Haecker), many languages make use of one or two words that are effectively untranslatable and in which is concentrated something of the genius of the language. The term *logos*, he says, is the main example of such a word in Greek. It can "be translated, according to context, 'word,' 'reason,' 'principle,' 'meaning'; but this fragments the connotation of the Greek word, which holds all these meanings together."[8]

Even in translation, however, the prologue of the Fourth Gospel hints at some of the nuances of the term. For, according to this prologue, it was not only through the Word—made flesh in Christ—that "all things came to be" in the beginning. In addition, "all that came to be was alive with his life" (John 1:1–4). Therefore, this passage not only posits an intrinsic link between the creation of the cosmos and what occurred historically in the person of Jesus, but it also gives a clear indication that, in some sense, the Word that "came into" the world in the person of Jesus had not previously been absent from it. And as Stephen Need has noted, this is precisely what a proper exegesis of the prologue suggests. "The incarnation in Jesus," he says, is not, for the Fourth Gospel, "the sudden arrival of an otherwise absent Logos, but rather the completion of a process already begun in God's act of creation."[9]

Presently, we will examine the background of this conclusion, together with the way in which its implicit panentheism was made more explicit by later Greek-speaking theologians. For the moment, however, we need to note only that a simplistic notion of the incarnation as a supernatural intrusion into the created order does not do justice to the subtlety of the biblical picture. This picture does not, therefore, directly challenge a naturalistic view of divine providence in the way it is sometimes thought to do.

This is not to say that taking the notion of the incarnation of God's *Logos* seriously does not modify the kind of strong theistic naturalism I have defended up to this point. This modification will, in fact, be the subject of a later chapter and, in many respects, the focus of my entire argument in the rest of this book. However, at this stage of our exploration, I shall not anticipate this argument in detail but simply indicate one of its aspects by noting an aspect of Christian belief that has often been associated with a strong incarnational stress. This is the recognition of what Arthur Peacocke calls "the sacramental view of matter."

What Peacocke seems to mean in his use of this term is summed up in his observation that there seems to be "a real convergence between the implications of the scientific perspective on the capabilities of matter and the sacramental view of matter that Christians have adopted."[10] God's creative action in and through "naturalistic" processes, he seems to suggest, may be compared directly with God's gracious action in the sacraments. When the cosmos is examined through the eyes of the scientist, Peacocke implies, there are clear echoes of those strands of Christian thinking that see the sacraments in terms of the transparency of created things to the divine purpose. Christians, he says, "starting, as it were, from one end of their experience of God . . . acting on the stuff of the world, have developed an insight into matter which is consonant with that which is now evoked by the scientific perspective working from matter towards persons, and beyond."[11]

Peacocke's recognition of the tendency of created things to be naturalistically oriented toward God's ultimate intentions is one that I have, in the light of these comments of his, elsewhere labeled as *pansacramentalism*. Although Peacocke himself may be reluctant to extend his own pansacramentalism in the direction of a strong theistic naturalism, such an extension is certainly possible. As all that I have said in this chapter indicates, the reasons that lead to this reluctance on Peacocke's part do not stand up to critical scrutiny. At the very least, these arguments suggest, a *pansacramental naturalism*[12] is not a position that can simply be dismissed as incoherent.

Therefore, until we come to develop a fully fledged *incarnational naturalism* toward the end of this book, I shall refer to the kind of strong theistic naturalism that I advocate as a pansacramental naturalism. In this way, the reader will be led to anticipate the more fully developed sacramental and incarnational insights through which my own version of a strong theistic naturalism will eventually be developed, and to remember that the sacramental insights that inform some of Peacocke's own views have had a significant influence in leading me to attempt this development.

5

The Acknowledgment of Miracles

When a certain kind of Christian conservative is confronted by the concept of a pansacramental or incarnational naturalism, that person is likely to object that, whatever the general implications of the doctrine of the incarnation, one of its traditional underpinnings—the biblical stories of Jesus' virginal conception—must surely count as a major stumbling block. For these stories recount a miracle, this critic will say, and a miracle (at least in common usage) is something that is in principle inexplicable in terms of the laws of nature. Therefore, according to this viewpoint, we can either affirm the historicity of this miracle or affirm that God acts always in accordance with "naturalistic" mechanisms. We cannot do both.

However, this assertion has less to do with anything that can be found unambiguously in the biblical or patristic literature than with later developments of Christian theology that led to the "God of the gaps" model. The basic meaning of the term *miracle*—"that which excites wonder"—makes no reference to "the supernatural" as such. Moreover, as we have seen, the pansacramental naturalism I have advocated does not make it necessary to declare that unless some phenomenon corresponds to our current understanding, it must be dismissed as simply impossible. For example, the concept of miraculous events that we have noted in John Polkinghorne's thinking—in which these events are analogous to regime change in the physical sciences—already means that our response to unusual or unique events must avoid any simplistic comment about certain kinds of events being evidently "contrary to the laws of nature." It is always possible that what is not straightforwardly repeatable may simply be a manifestation of extreme conditions that we do not yet fully understand.

The issue of reported incidents that are neither understood nor straightforwardly repeatable is, in fact, philosophically complex. No scientist, for example, believes that we already understand all that there is to be understood. The fact that some event is at present inexplicable says no more in itself, therefore, than that our current understanding is incomplete. What is more, the question of observational or experimental repeatability is by no means as simple as some take it to be. Repeatability of experiment or observation is, admittedly, of the essence of the scientific method. Nevertheless, it is clear that certain kinds of phenomena—of the sort that scientists accept as aspects of robust theoretical frameworks—are in practice unrepeatable because of the complexity of the situation that is necessary for their occurrence. For example, the fact that the emergence of a particular species of animal is not in practice a repeatable phenomenon rightly makes no dent in the scientific consensus about how such emergence occurs.

The issues that arise from this for our consideration of miracles are illustrated in an interesting way by the current confused status of research into "paranormal" phenomena. Important questions arise from the fact that repeatable experimental demonstration of such phenomena (telepathy, telekinesis, and the like) has proved impossible despite more than a century of research. Many people, in the face of this fact, simply dismiss the possibility of such phenomena, putting all anecdotal evidence of their occurrence down to fraud, wishful thinking, or pure coincidence. Others, however, consider that the strength of the anecdotal evidence is such as to suggest that phenomena of this sort do actually occur but are extremely difficult to investigate through experimental methods.

If we take seriously the way in which, even in well-established scientific disciplines, practicable repeatability decreases with increasing complexity of the relevant causal factors, then we shall perhaps see this latter approach as the more coherent one. Moreover, once we recognize that high degrees of complexity may well characterize the sort of phenomena that are labeled paranormal, we shall necessarily come to a conclusion different from that of the utter skeptics. We shall not only recognize that holistic top-down organizing principles—analogous to those used to refute reductionism—may be at work in paranormal phenomena (which in this sense may be seen as manifestations of yet another "new emergent" property in the cosmos). We shall also acknowledge that analysis of anecdotal evidence of the occurrence of such phenomena must be extremely complex, not simply a matter of asking whether what has been reported is repeatable in the laboratory.

In recent philosophical discussion, this issue of the complexity of causal factors has been examined less in terms of questions about paranormal phenomena than in terms of the more general question of what it means to speak about causality at all.[1] One aspect of this is illustrated rather nicely by a simple example that is sometimes used in this discussion: that of how different observers may have quite different perspectives on the "cause" of some event. In the case of a forest fire, for example, a human observer is likely to speculate about possible causes in terms of prior events, such as a lightning strike or a carelessly discarded match. But from the viewpoint of the hypothetical passengers of a Venusian spaceship flying overhead, the cause may seem quite obvious and will be quite different from the human speculation. "What can you expect," they will ask, "when the earth's atmosphere has so much oxygen in it?"

The point here is that levels of explanation are needed for a complete analysis of causality. The cause of any event is never simply a prior event. Only a full knowledge of the total context in which any event occurs allows a complete explanation of its occurrence. The human observers of the forest fire look for explanation at a "low" level that takes for granted the possibility of burning. The Venusians perceive a "higher" level of causality, which the humans tend to ignore and of which they may be unaware. Yet other observers may perceive the working of other "higher" factors of this kind—say, ones related to the chemical bonds of combustible materials, which make the chemical reactions involved in their response to heat exothermic rather than endothermic.

This point about causality was, interestingly, implicitly recognized in the patristic period. In Augustine's framework for the discussion of miracles, there is a clear implication that highly unusual events are able to occur because there is, over and above the natural law that we are able to understand, a "higher" lawlike framework that the cosmos also obeys but that is in practice beyond our understanding. If there are simple systems that are susceptible to human understanding in terms of the "lower" law, this is, Augustine seems to suggest, only because the threshold has not yet been reached at which the influences of this "higher" component of natural law are significant in their effects.[2]

The pansacramental naturalism that I advocate can clearly be developed by acknowledging, in a broadly similar way, the possibility of events that, while repeatable in principle, cannot be straightforwardly reproduced, because the conditions that are necessary for their occurrence are either unknown or cannot, practically, be replicated. In this way, a strong theistic naturalism need not deny the historicity of reported events simply because they are currently (and perhaps inevitably) inexplicable. Rather, by invoking either the regime change analogy or holistic top-down organizing principles, this naturalism can acknowledge such occurrences as being in principle possible, even if in practice extremely difficult to authenticate.

This question of authentication is, in fact, the crux of the matter. Methodologically, it is arguable that acknowledgment that some paranormal event has occurred can only be given after exhaustive considerations of several kinds. In particular, the history of paranormal research indicates the prevalence of fraud or wishful thinking in anecdotal evidence of unusual events, and there seems little reason to doubt that a similar prevalence is a feature of comparable reports that have been interpreted in a religious way. (Indeed, the element of wishful thinking may be even stronger.) Acknowledgment of probable authenticity must therefore require that some particular reported incident be supported either by very strong anecdotal evidence—a large number of witnesses, for example—or else by wider considerations that are judged to be relevant and of considerable significance.

What these wider considerations should be in a theological context can be illustrated in part, perhaps, by the way in which the Fourth Gospel speaks of the miraculous events that it reports as "signs." These, as commentators have long recognized, have an explicitly theological function in terms of conveying a spiritual symbolism, which is brought out in the gospel by the discourses with which they are interwoven in the text.[3] In general terms, we can perhaps best express this "sign" aspect in terms of an existential deepening of understanding and commitment. In this perspective, what counts as a miracle or sign is not necessarily inexplicable in naturalistic terms. Rather, a miracle is what Ian Ramsey has called an event that—while perhaps explicable at one level in the language of natural laws—requires also for a full understanding "an odder language which witnesses to the fact that the situation is *only in part* perceptual." What characterizes a true miracle is the evocation of this "odder language": the necessity to express "what is distinctively religious *in personal terms*."[4]

In the context of an essentially naturalistic understanding of the sort I have advocated, the implications of these considerations might be illustrated by a somewhat unreal example. Suppose I experienced the spontaneous boiling of the water in an unheated kettle, just at a moment when I had been longing for a cup of tea and (having no fuel for the stove) had prayed to God about the problem. What would be the response of those to whom I reported this event?

Perhaps most listeners would deny that it had happened, on the grounds that it was "impossible." Others might accept that it had happened and interpret it as a supernatural miracle. The physicists to whom I reported the event would, however, point out that such an occurrence as the spontaneous boiling of water is not actually impossible. (Spontaneous boiling is, in the perspective of modern physics, merely highly improbable, even if we wait for billions of years for it to happen.) These physicists would therefore make a judgment of my report that was based partly on their calculation of the event's actual probability and partly on their estimate of my reliability as a witness.

The atheists among these physicists would, of course, consider the prayer that had preceded the spontaneous boiling as no more than a coincidence, comparable to that which might occur in the case of a lottery winner who had previously prayed to God about financial problems. The win was highly unlikely for any particular individual, they would say, but it was bound to happen to one of the millions who had taken part in the lottery, and the prayer of the actual winner was irrelevant.

The theists among them might see things differently. Those who believed in special providence would suggest that it was not intrinsically impossible that God had brought about this otherwise unlikely event, either supernaturally or through some sort of causal-joint mechanism. They would therefore ask, not "Was this occurrence possible?" but rather, "Was this the sort of situation in which God would be likely to bring about an event of this sort?" Those who were strong theistic naturalists would, in a slightly different way, ask essentially the same question. They would ask, not "Was this occurrence possible?" but rather, "Was the total context here such that the fixed instructions that God has built into the world would give rise to an event of this kind?"

Thus, for both classes of theistic physicists, the main question would not be that of possibility, but rather would be whether what was reported conforms to the true definition of a miracle: that which acts as a sign of God's revelation or will. In the particular case of the spontaneous boiling of my water, most of them would say that the answer to this question is no, since miracles—however understood—are not usually thought of as appropriate in cases of longing for a cup of tea. It is, in fact, precisely the lack of any "sign" element that makes this particular example as absurd as it is.

In most instances of reported miracles, the question of their naturalistic probability will not be as well defined as in this hypothetical example. Nevertheless, I would argue, our actual ignorance of what is possible in the natural world—of what in quasi-Augustinian terms we can call the "higher law"—is still so great that we cannot respond to a reported miracle simply by claiming that it could not have happened.

For whether we are strong theistic naturalists or believers in special providence, if we take the providential experience of the Christian community seriously, we necessarily believe in what John Polkinghorne calls a "supple and subtle" universe, in which extremely unusual things can happen. Therefore, when we are faced with reports of a miracle, a degree of skepticism may still be methodologically justifiable, but our main questions should be about whether the anecdotal evidence of its occurrence can be seen as weighty, and about whether and how it acts as a sign.

We will not always be able to give a straightforward answer to either question, of course. For example, in the case with which we began this chapter—that of Jesus' reported virginal conception—there are clearly important historical and theological arguments on both sides. At the historical level, skeptics about the basic historicity of the virginal conception can legitimately ask nonskeptics how they can account for the many differences between the two gospel passages that report this conception, and for the absence of any obvious knowledge of a virginal-conception tradition in the other two gospels and in the Pauline epistles. Conversely, these skeptics must themselves face the difficult question of how a virginal-conception tradition, if it were historically spurious, could have arisen at a very early stage of the Christian community's history, and they must recognize that their usual assertion—that the accounts arose as christologically inspired glosses—is itself problematic as a historical scenario.[5]

At the theological level, there might be no doubt that the virginal-conception accounts act as a "sign." However, since theological interpretation has clearly been an important factor in the redaction of the Gospels and in the development of the oral tradition on which they are based, the relationship between this "sign" and the historicity of the story that communicates it is difficult to unravel. Moreover, not only—as one defender of the accounts' historicity has noted—are "the doctrines of the sanctity of Mary and of the incarnation of God's son . . . not logically dependent on the virginal conception,"[6] but it is also arguable that the most radical alternatives to such a conception are not themselves without sign-bearing potential.[7]

For a number of reasons, then, this issue is complex. Skeptics may acknowledge the conception stories as a psychologically potent component of the Christian proclamation[8] but feel able to interpret this in terms of the importance of mythology to the instrumental effectiveness of the Christian story. Following the example of the author of the Fourth Gospel, they may feel no incongruity in proclaiming that "the Word became flesh" in Jesus while at the same time feeling it unnecessary to make any historical claims about the way in which he was conceived. Conservative Christians, in contrast, will tend to see this judgment as flawed and may well claim there are further factors to be taken into account. If they are Protestant fundamentalists, for example, they may invoke the "infallibility" of the Bible; if they are Roman Catholics or Eastern Orthodox, they may invoke the importance of the role of Jesus' mother in the economy of God.

Here, rather than elaborating on my own response to this debate,[9] I want to stress that my pansacramental naturalism does not, in itself, favor one side or the other. If, as is arguable, questions about the historicity of Jesus' virginal conception may legitimately be asked, then this can only be because weighty historical and theological considerations give rise to such questioning. What such questioning cannot call to its aid, however, is any form of naturalism other than the rather shallow version of it that usually prevails, in which virginal conception is simply deemed impossible. The pansacramental naturalism that I advocate here is not, it must be emphasized, a naturalism of this shallow kind.

Indeed, I would go further. Whatever our judgment may be about any particular reported "miracle" (and the reasons for skepticism may often be great), the total religious experience of humankind suggests to me the need to be open to the possibility that phenomena of the kind usually deemed miraculous do occur. Such occurrences would not, for the model that I have outlined, mean that the logic of the universe's functioning has been violated, but simply that this logic may not be completely comprehended by the human logical faculty. (The universe's logical nature is, after all, grounded for the Christian not in that faculty but in the divine *Logos* itself.) There may be laws of nature about which we know nothing scientifically but that nevertheless occasionally have significant effects.

In its implications for our understanding of divine action, my view here is comparable, in philosophical terms, both to that of John Polkinghorne, when he speaks of miracles as being analogous to changes of regime in the physical world, and to that of Robert John Russell, who talks of the possibility of a unique event of religious significance being a "first instantiation of a new law of nature."[10] There are, I believe, some laws of nature that can—if only under very unusual circumstances—bring into effect what we can describe theologically as a realization of the world's eschatological potential.

This realization is what Christians sometimes call a "breaking in" of the age to come. For my model, however, this phrase is somewhat unfortunate, since what I envisage is not a "breaking in" of something that comes from "outside." Rather, as I shall explain in more detail presently, what I envisage is something that the Eastern Christian tradition has often stressed: a "breaking out" of something that is always present in the world, albeit in a way that is usually hidden from us. If this hidden aspect of the world is not susceptible to scientific investigation, this is not, I shall suggest, because it is not susceptible to a naturalistic understanding. Rather, it is because its manifestation depends on something that cannot be replicated under laboratory conditions: the faithful response to God of those who recognize him as their creator and redeemer.

6

Risen and Ascended

For many conservative Christians, the Gospels' accounts of the finding of the empty tomb and of encounters with the risen Christ have had much the same function as the accounts of Jesus' virginal conception. When accepted as historically accurate (at least in broad terms), both sets of accounts have been widely taken as guarantees that God has, in the person of Jesus, acted in the world in a supernatural way.

As we have seen, however, nothing that a supernaturalist framework affirms is intrinsically impossible within the sort of pansacramental naturalism I have outlined. Acceptance of that naturalism does not, therefore, make it necessary to deny the basic historicity of the empty tomb and resurrection appearances. If we are to question the historicity of these accounts, then—as with the virginal-conception accounts—it must be on grounds other than the assumption that "things of that sort cannot happen."

The potential for such questioning clearly does exist, however, since the accounts of the resurrection appearances have proved far from immune to skepticism based on historical-critical analysis. Nevertheless, there are several reasons to affirm at least the basic historicity of experiences of the risen Lord. One is that a resurrection tradition, unlike a virginal-conception tradition, is widely accepted by historians as having been universal in the earliest strands of Christian proclamation. As a result, it is difficult to claim that it was merely a late and essentially mythological gloss on the earliest Christian belief. Moreover, even if we do make this claim, as one minority strand of New Testament scholarship does,[1] there is still the need to account for the astonishing transformation of Jesus' followers in the period following his death. The conversion of these followers from a frightened and defeated group into the core of a confident and death-defying church is extremely hard to explain unless we posit some significant experience of the risen Lord whom they went on to proclaim.

Although this kind of reasoning does point to the historicity of some kind of potent experience of the risen Lord, it does not provide grounds for belief in the historical reliability of the details of the biblical accounts. Therefore, we are still left with the question of how we should deal with the evident elements of truth in the observation that the mutual incompatibilities of these accounts indicate that complex oral and redactional processes lie behind them.

One possible way of dealing with this problem is to acknowledge that the various "skeptical" attempts to understand the nature of the Easter experiences may, in some respects, actually be helpful. Take, for example, the view of Norman Perrin and others: that the best insight into the nature of the original experiences may, in practice, not be through the gospel accounts but through the conversion experience of the apostle Paul.[2] To some, this approach may seem entirely unacceptable because the visionary nature of the Pauline experience inevitably leads to questions about these earlier experiences in terms of human psychology. Such questions cannot, however, be avoided simply by adopting a more conservative judgment about the subjective content of the Easter experiences, for there are (as I have pointed out elsewhere[3]) striking parallels between the experiences recorded in the biblical accounts and certain types of secular visionary experience. Thus, even if we ask no questions about the basic historicity of these accounts, we still face the questions of whether and how the reported experiences might be analyzed in a psychological framework.

The question of whether the earliest experiences of the risen Christ were essentially visionary does not, however, arise primarily from approaches such as these. It arises from the fact that the biblical accounts themselves indicate that only believers saw the risen Lord, that his entrances and exits were mysterious, and that he was sometimes unrecognizable at first. These factors clearly suggest that, whatever the witnesses to Christ's risen state experienced, it was not something with the attributes of an ordinary object. It is hardly surprising, therefore, that "theories of visions" constitute a long-standing aspect of theological discussion of the narratives.

Over three-quarters of a century ago, for example, Edward Gordon Selwyn could write of these theories as being already of considerable age and falling broadly into two categories. In one set of theories, he noted, the visions were perceived as subjective: "simply a product of the disciples' mental condition at the time." In the other, they were declared to be objective, in the sense of having been "caused by the invisible Christ Himself, really present with them. The disciples were inspired by God to see what they saw: Jesus was really alive, and the eye of faith could behold Him."[4]

Selwyn himself was perceptive enough to challenge this kind of categorization as simplistic and to recognize a more complex spectrum of possible views. Both of these kinds of theory are nevertheless still with us. Some try to explain away the beginnings of the Christian community in terms of subjective hysteria. Others insist, by contrast, that the resurrection was an objective event. God, they assert, chose to reveal this event through the disciples' imaginative faculties because a revelation of the fullness of the glory of the risen Christ was beyond what they could otherwise grasp.

A classic twentieth-century exposition of this latter view is that of Hans Urs von Balthasar, who (like Selwyn) brings to bear on this issue his familiarity with the Western mystical tradition's understanding of visions. The resurrection appearances can, says von Balthasar, be understood in visionary terms because all authentic mystical visions involve the presentation of an objective reality in "a form that appropriately expresses and reveals it." This kind of presentation need not, in his view, bypass all natural psychological processes, because God can be revealed objectively through processes that "bring into play the mystic's own imagination."[5]

In a similar way, Karl Rahner has also drawn on the Western mystical tradition to analyze revelatory visions in terms of the imaginative faculty. In his analysis, however, there is at least a slight move away from a purely "objectivist" understanding. Noting that the visionary experiences reported throughout the Bible have, like those of the church's mystics, manifested great variety, he suggests not only that such experiences are usually of "created symbols," which vary widely "according to the phase of history of salvation in which the visionary lives and which he is intended to influence," but also that a vision "may be conditioned in part by the historical milieu in which it occurs."[6]

Addressing what he calls the psychological problem of visions, Rahner notes that the writings of the mystics of the Western church traditionally categorize visions into three kinds: the corporeal, the imaginative, and the purely spiritual. Of these, he notes, the mystics themselves "regard the imaginative type as the more valuable and exalted," and he argues from this and from a number of other observations that "the 'authenticity' of a vision cannot be simply equated with its corporeality, its objective presence within the normal sphere of perception affecting the external human senses."[7] The problem of interpreting any vision of a person as a literal encounter with that person's real or glorified body is considerable, he notes, and there is no significance in the fact that the person seen may have given "an impression of reality" by moving, allowing himself—or herself—to be touched, or imparting information.[8]

Even in cases of more than one person experiencing a vision—some of which Rahner clearly considers authentic—he sees good reason to be skeptical of their corporeal nature. Noting that hallucinations are sometimes experienced by a number of people simultaneously, he simply recognizes such multiple experience as an aspect of human psychology (albeit ill understood) and notes that if "a psychic mechanism can be started simultaneously in a number of people, then the possibility that God too might make simultaneous use of these psychic potentialities in a number of people cannot be rejected *a priori*."[9]

Rahner's conclusion—based on a number of lines of reasoning and in no way based on a reductionistic skepticism about the authenticity of religious visions—is that in most cases authentic visions will be imaginative ones. Moreover, while he is careful, with his Western Catholic grounding, to affirm that God *can* suspend natural laws and produce visions that are miraculous in the sense of having an interventionist, supernatural input, he goes on to qualify this in a fascinating way: "Even where a miracle really occurs, we can consider the laws of nature suspended only to the extent strictly required by the event; and in such cases as far as it is possible God will use the natural laws which after all he has created and willed." Thus, Rahner goes on, "it is to be expected that even in visions of divine origin the seer's psychical structure and the laws of his nature will remain intact and operative to the fullest extent."[10] To affirm that a vision is the work of God, he says, we do not have to assume supernatural input.

Rahner also goes on to distinguish two categories of possible supernatural input. In addition to the technical miracle, involving a suspension of the laws of nature (including the normal laws of psychology), there is the type of act that involves what is sometimes, in Western Catholic terminology, called "sanctifying grace." This, Rahner notes, is a "causality of God . . . which transforms nature but does not suspend its laws in any proper and empirically verifiable sense."[11] The technically miraculous, he insists, "need by no means be more perfect, either ontologically or ethically," than that which comes about through natural laws. Moreover, he argues, "visions involving a partial or total suspension of natural laws must not on that account be considered 'corporeal' visions. They too can be . . . purely imaginative ones."[12]

Rahner emphasizes, then, that authentic religious visions are usually imaginative rather than corporeal and—by implication at least—are usually either entirely "natural," in the sense of being the product of the human psyche, or else supernatural only in the strictly technical sense of involving not an intervention in natural psychological laws, but rather the kind of divine influence that allows those laws to become unusually transparent to the divine purpose. (This view, as we shall see, can be expressed straightforwardly in terms of the pansacramental naturalism that I advocate.)

We cannot, of course, ignore the fact that Rahner, unlike von Balthasar, does not explicitly apply his insights into visionary experience to the Easter experiences themselves. The reasons for this seem to lie at least partly in his apologetic motivation. In his work, not only—as one commentator has put it—are "historical details and hermeneutic fine points . . . subordinated to the one over-riding concern, namely the demonstration of the possibility of being a Christian today without offence to one's rational nature."[13] But also, as another has noted, Rahner often seems to be deliberately avoiding explicit historical application of his insights because of his fear of the "consequences of free floating research into the circumstances and intent of texts etc."[14]

The argument for extrapolating Rahner's understanding of secondary revelatory experiences to primary ones does not, however, rest simply on recognition of his apologetically motivated reticence. It is also arguable that an extrapolation of this sort is implicit in much of his work. For, as Christopher Schiavone has noted, a central aspect of Rahner's framework is his belief that God's action on human beings is primarily at a "contemplative" level that is deeper than that of either sensation or rational thought. In this way, says Schiavone, Rahner "locates 'primitive revelation' (i.e. revelation from the Patriarchs to Jesus) in a broader history of revelation as a whole."[15]

Once we recognize that this contemplative dimension is, for Rahner, the very locus of God's revelatory activity, we can see that his analysis of secondary visions has at its heart an affirmation that applies equally to primary visions. This affirmation is that "even in the imaginative vision it is not as a rule the vision as such (the stimulation of the sense organs) that is primarily and directly affected by God. Rather the vision is a kind of overflow and echo of a much more intimate and spiritual process . . . which the classic Spanish mystics describe as 'infused contemplation'. "[16]

This stress on the contemplative dimension of revelatory experience seems to provide a theological justification for analyzing the Easter experiences of Jesus' disciples in visionary terms, and this provides an important supplement to the historical grounds for speaking of those experiences in this way. As we have noted, this does not, in itself, demand that the visions must be seen as purely subjective; indeed, it is arguable that even when expressed in naturalistic terms, this understanding permits maintaining an essentially "objective" understanding of the contents of the resurrection appearance visions.

But how strongly do we need to insist on the "objectivity" of their contents? Certainly, as Christians, we want to insist on a genuinely referential aspect of what has been revealed to us through the appearances of the risen Lord reported in the New Testament. Equally, however, certain aspects of these experiences are problematic for us. Take, for example, the last of the resurrection appearances as reported by Luke: that which culminated in the ascension (Acts 1:1–10). How are we to understand this account when we can no longer regard heaven as a "place" to which one can literally ascend? Must we deny the historicity of the reported experience? Or can we affirm its historicity in a way that recognizes that what was experienced could not have been a literal, spatial "ascent"?

This latter option will be open to us only if we can find a way of affirming that the experience manifested an "objective" reality through something that was less than absolutely objective. It is precisely here that Rahner's understanding of revelatory visions—as "a kind of overflow and echo of a much more intimate and spiritual process"—becomes extremely important. For it enables us to affirm the absolute authenticity of the "inward," contemplative element of any valid revelatory vision, while at the same time recognizing that the "outward" contents of that vision may reflect conditioning by what Rahner calls its "historical milieu."

What this means in practice can be indicated by the way in which, in the creeds of the church, an affirmation of Christ's ascension into heaven is immediately followed by an affirmation that he "is seated at the right hand of the Father." What makes these assertions somewhat curious for us is that we cannot accept either of them as having any kind of spatial reference. Just as we do not think of heaven as a "place" to which one can literally make an "ascent," so also we do not see God as having something akin to a physical form,[17] which might allow us to talk about God having a right, as opposed to a left, hand. Because of this, we assume that the notion of Christ sitting at the right hand of the Father is essentially a poetical pointer toward the way in which Christ may be seen as having a status and function analogous to that of the grand vizier, who sits at the right hand of an oriental potentate and exercises authority in his name.

Therefore, when we read of the first martyr Stephen seeing this vizier status in visionary form (Acts 7:56), we understand this vision as involving a created symbol of great psychological power. We do not deny to it a genuine reference, but we see this reference as being to something far more complex than a spatial relationship. In a similar way, when it comes to the notion of Christ's ascension into heaven, we cannot interpret it, as the earliest Christians tended to, in terms of a literal bodily ascent to a "place" above the sky. In the context of Rahner's insights, we can, however, interpret the account of the ascension experience in terms of a visionary experience comparable to that vouchsafed to Stephen. The usual "theological interpretation" understanding of the account—in which its basic historicity is denied[18]—is not the only one open to us.

On its own, perhaps, this insight is inconclusive in relation to the question of whether the ascension account in Acts is best interpreted in terms of historical memory or of theological interpretation. However, a further factor for our consideration arises from the observation that the theme of ascension was widely used in the Judaism of the time.[19] The importance of this observation is indicated by an aspect of the resurrection appearance accounts that is only rarely discussed in any detail: the way in which, in Matthew's gospel, it is reported that "many of God's people arose from sleep and, coming out of their graves after his resurrection, entered the holy city where many saw them" (Matt 27:53).

The relevance of this passage lies in the observation that what it expresses is far less easily explicable in terms of "theological interpretation" than is the ascension account in Acts. Among the early Christians, there was no questioning of the notion that Christ had ascended into heaven, but this account of others who had risen from their graves was much more difficult to fit into the accepted picture of the prevailing state of things.

Faced with the question of whether we should see this Matthean report as a genuine reflection of the content of the earliest resurrection experiences, many commentators answer in the negative. Their reasons for this are, however, essentially the same as those they cite for denying that there was a historical ascension event that left the disciples "gazing intently into the sky" (Acts 1:10). Just as this latter account clearly incorporates a culturally conditioned mistake about the "place" of heaven, so the account in Matthew, it is rightly noted, incorporates a culturally conditioned mistake about the scope of the resurrection. It reflects the widespread expectation, before Jesus' death, that the resurrection would not be of one man only, but of all just people.[20]

This, however, is precisely the point. While the notion that Jesus had literally made an ascent into heaven was still believable at the time when the evangelists were writing, the notion that the general resurrection of all just people had already occurred had been rendered void. The general resurrection of the just was not, for Christians of the late apostolic era, something that had already occurred. It was still to come, at the time when Christ himself returned in glory. The "mistake" was, so to speak, a tenable one in the immediate aftermath of Jesus' death, but not in this later period.

This fact points strongly to the way in which cultural expectations are likely to have significantly affected the content of the experiences of the resurrection appearances, and this underlines and illuminates Rahner's point about the way in which any revelatory experience may be "conditioned in part by the historical milieu in which it occurs." Just as it would seem to make more sense to see Matthew's report of the experience of the general resurrection of the dead as reflecting a genuine and very early historical experience, so also, I would argue, we should see the ascension experience as being rooted in the historical memory of a culturally conditioned vision, not simply as later theological interpretation.

The usual objection to this conclusion is, of course, that such an experience was "impossible." A transcendence of this apparent impossibility is, however, precisely what the hypothesis of culturally conditioned visionary experience allows. For the common assumption that certain biblical passages can have no basis in historical memory is challenged radically by a visionary model of revelatory experience, in which what is experienced historically does not have to correspond to what is possible empirically. The model allows us to affirm the absolute authenticity of the "inward," contemplative element of the experience, while at the same time recognizing the way in which its "outward" contents may reflect the expectations and understanding of the period.

In this way, the hypothesis of the culturally conditioned vision has a remarkably conservative aspect, for it allows us to see genuine historical memory in several accounts to which it has commonly been denied. Because of this, it can be extremely helpful to Christians who, without resorting to a fundamentalist denial of the validity of the questions posed by historical-critical analysis, want to affirm both the basic historicity of the resurrection appearance accounts and the traditional theological understanding of their meaning.

7

Revelation and Salvation

The visionary and contemplative understanding of revelatory experience that I have outlined has ramifications that are, as we have seen, extremely helpful to conservative but nonfundamentalist Christians. It allows a way of affirming the historicity of the resurrection appearances reported in the New Testament without having to insist that the critical questions that immediately occur to us in relation to these accounts are invalid. However, this understanding has other ramifications that may, to some, seem more worrying. For it clearly indicates that what was experienced in the aftermath of Jesus' death was not "objective" in any straightforward sense of the term, but was, at the very least, sometimes colored by the cultural expectations about the end of the age that characterized the Judaism of the period. The model suggests, therefore, that we may need to make an important distinction between different aspects of revelatory experience.

On the one hand, the model suggests, any authentic revelatory experience will have aspects that are genuinely referential to a reality that God wishes to make known to us. On the other hand, the experience may also have aspects that simply reflect the kind of cultural expectations that have made the psychological experiences possible. In this sense, the model represents what I have elsewhere called a "psychological-referential model of revelatory experience,"[1] in which an important distinction must be made between what is referential in any revelatory experience and what represents simply an aspect of the psychological vehicle through which the experience has come about.

One way of approaching this distinction between reference and psychological vehicle is provided by the aspects of the Jungian understanding of human psychology that claim to throw light on the nature of religious experience. For this understanding, such experience is always related to the "archetypes of the collective unconscious" that, according to Jung, reveal themselves in the symbolism of all the religions of the world. Religious experience, for this understanding, always involves an essentially naturalistic psychic process in which these universal archetypes manifest themselves at a conscious level, usually in a way that promotes psychic health.

This approach is of considerable interest here, since it can, for those who accept it, provide a secular underpinning to the sort of psychological-referential understanding I have outlined. In particular, it can provide an illuminating way of thinking about both the universality of religious experience and the cultural specificity of its actual manifestation. For while Jung stresses that the archetypes of which he speaks are present in all people, he also notes the way in which the form of their eruption into consciousness may be determined very specifically by a particular cultural and historical context.[2] Moreover, the naturalistic element of this understanding is particularly relevant here, because it points to the way in which the kind of psychological-referential understanding that has arisen from my reflection on revelatory experience may be expanded in terms of the sort of pansacramental naturalism that has arisen from my more general considerations about divine action. In this sense, Jungian insights can provide an important underpinning to any attempt at such an expansion.

It is important to recognize, however, that a naturalistic interpretation of the psychological-referential model of revelatory experience does not depend on any particular judgment about the Jungian (or any other) psychological scheme. Such an interpretation arises primarily through a creative combination of my pansacramental naturalism, with its roots in theological and philosophical analysis of divine action, and my model of revelatory experience, which arises primarily from historical and philosophical reflection on God's self-revelation in the person of Jesus. Jungian insights are, in this context, no more than peripheral. They suggest a secular approach to psychology that may prove helpful in examining some aspects of this combination. They are not, however, an integral part of it.

Recognition of this is important because appropriation of the Jungian framework is theologically problematic. It may be true that the most common theological objection to this framework is itself questionable, since it is based—as Jung himself notes—on a denigration of the human psyche that arises from a stress on God's transcendence at the expense of any appreciation of God's immanence.[3] Even when we acknowledge this fully, however, the Jungian framework still cannot be integrated into a theological perspective without major modification or expansion. Not only may many questions be asked about it in its own quasi-scientific terms,[4] but more importantly, Jung's framework gives rise to fundamental questions that he himself tends to answer in a tendentious way.[5] Indeed, the main theological issue that arises from my psychological-referential model of revelatory experience is one that Jung studiously avoids: the question of how we can distinguish what is referential in any such experience from what is no more than psychologically instrumental. Arguably, this distinction can be made within a theologically expanded Jungian framework.[6] It must be emphasized, however, that such a distinction does not in any way depend upon an expansion of this kind. It must rely primarily not on psychological considerations but on theological ones.

Sometimes distinguishing between referential content and psychological vehicle in any particular experience may seem to us a simple enough theological task. For example, in the case of the ascension account in the Acts of the Apostles, few of us are likely to find it easier to accept a literalist reading of the reported experience than to accept a visionary explanation that makes a distinction of this kind. The reference in this experience, we are immediately tempted to suggest, was to Christ's transformation from a partially "local" manifestation to more generally cosmic presence, while the psychological vehicle was a vision of his body undergoing an ascent into the clouds, which provided the necessary means for this transformation to be understood and responded to in faith.

To make the distinction in this way is, however, questionable, for by using the psychological-referential model to interpret the ascension in this way, we have made implicit judgments that are not intrinsic to the model itself. Why, we must ask ourselves, have we judged the referential component of the ascension experience as we have? Might there be aspects of this experience that we quasi-instinctively take to be referential but that in fact should be assigned to the other side of the divide?

This issue is illustrated well when we consider the resurrection appearances in general. In the context of the first century, when an expectation of resurrection at the end of the age was an important aspect of some strands of Judaism, we must ask ourselves whether, when we attempt to understand the resurrection experiences, we should consider only the visions of the ascension and of the general resurrection of the just to have a questionable referential status. When we recall, for example, von Balthasar's and Rahner's comments about imaginative religious visions being at least as valid as corporeal ones, we may wonder how we should understand the experience of Christ's "resurrection body." Should we see this aspect of the experience of the risen Christ as being referential to an objective aspect of Christ's new being, or should we see it simply as a component of the psychological vehicle for the communication of the reality of his "spiritual" resurrection?

However we may choose to answer this particular question,[7] we can see it as an example of the way in which, in a psychological-referential theory, we seem forced to recognize the possibility that any referential content in revelatory experiences may, to some extent, be "hidden" within the culturally influenced expectations in and through which those experiences are psychologically appropriated. This has the corollary that the doctrine that is initially "read off" from the experiences will not necessarily be straightforwardly referential. Rather, whatever their salutary instrumentalist effect at an unconscious level, the perceived contents of revelatory experiences will be at best only *candidates* for truly referential doctrine.

A psychological-referential model of revelatory experience thus presents us with an apparent dilemma. On the one hand, as we have seen, it provides a plausible solution to some of the puzzles that arise in relation to certain historical experiences. On the other hand, while the model can affirm that these foundational revelatory experiences have provided us with real, referential knowledge of God, it also acknowledges the difficulty of distinguishing this referential knowledge from what was simply the psychological vehicle by which it was conveyed.

One way in which we might approach this dilemma is through the recognition that our inner conviction of the referential validity of certain religious doctrines does not rely on a simplistic interpretation of the revelatory experiences from which those doctrines have been "read off." Our Christian faith is tied also to our confidence that its doctrinal framework makes sense of our total experience of the world. What underpins our affirmation of the referential nature of at least some doctrinal statements is their consonance with experience: their way of appearing to solve some of the existential puzzles inherent in our existence.[8]

This assignment of aspects of revelatory experience to the referential category will in practice, however, often be quasi-instinctive rather than logically unassailable; as a result, it can only be provisional. I would argue, however, that this need not be worrying, provided that we can see the importance of the notion—common to many strands of recent theological thinking[9]—that God's self-revelation is not primarily a matter of conveying "information" at all. Rather, the concept of revelation may be better understood when it is considered in terms of what is needed for our spiritual development, with conveyance of referential information as only a secondary aspect.

This sense of the soteriological dimension of revelation may have been best expressed in Christian terms in the work of Yves Congar, who stresses that we will know God as fully as is possible for us only when this present age ends, at the *eschaton*. In the present age, he says, God's self is made known to us, not in abstract knowledge, but in "signs," which are always oriented primarily toward salvation, "being proportionate to our human condition, and couched in the language of men, in images, concepts, and judgments like our own."[10]

Congar does not deny these signs a genuine ontological content. He does, however, insist that this content is conveyed in terms of "mysteries," which are partially hidden truths, made present most fully in the liturgical celebration of salvation, and which always, when expressed linguistically, are to be understood apophatically. Since all revelation is, in Congar's view, of God's self, and tending toward a final, eschatological fruition, he talks of an "economy of the disclosures of God" in terms of three "unveilings of our eyes." The first of these relates to what may be known of God by reason, the second to what may be known by faith in response to God's self-revelation in history, and the third to the eschatological vision of the blessed.

As William Henn has observed, not only does this perspective "fit nicely into Congar's historical approach to truth," but in addition, "the view that revelation is progressive and will be completed only at the eschaton serves as a major limiting principle with regard to the adequacy of any present statements which intend to convey revealed truth. All such statements, in principle and by their very nature, must be treated in a sober manner, which is aware of their 'proleptic,' provisional and not-fully-adequate nature."[11]

This kind of understanding is, I believe, important in several ways. It clarifies an aspect of the sort of traditional apophaticism that we have noted, and it also complements the important concept of liturgical action—which we shall consider presently—in which that action actually "makes present" the central realities of the events it "commemorates," allowing them to be appropriated by believers in a way that transcends their intellectual understanding. Here, however, its importance lies primarily not in these factors, but in the way it may be expanded in terms of a pansacramental naturalism of the sort I have outlined.

As we have seen, such a naturalism holds that all created things are oriented toward their ultimate end in Christ. In the case of the least complex aspects of the creation, this can be seen in the way in which the cosmos is inherently fruitful; it has, through naturalistic processes, become ever more complex and has given rise to new emergent properties. The result has been the existence of ourselves: beings with intelligent self-consciousness who can make the free choice of responding to God or of failing to do so.

A pansacramental naturalism of this kind cannot, however, be limited to this level of development. If it were, then it would not in fact be oriented toward eschatological consummation. At this level, the ultimate end is nothing more than the death of the universe (which, depending on the cosmological model we adopt, will be either the "heat death" of an ever-expanding universe or else the "big crunch" of one that eventually collapses due to its own gravity). If, therefore, a pansacramental naturalism rightly focuses initially on aspects of the cosmos that allow the emergence of beings who can respond to God in faith, it must also focus on the ultimate divine purposes that are indicated by the potential for this response. This understanding points beyond the end of the physical universe to what the Christian describes as the *eschaton*.

Here the psychological-referential model of revelatory experience that we have examined becomes extremely significant. For when we interpret this model in terms of a pansacramental naturalism, we can see how and why human psychology—an emergent property of the cosmos—may be seen as having a truly eschatological orientation. Our human nature may be seen as being capable, in its psychological dimension, of knowing God, not in terms of abstract "information" about God, but in terms of the deeper kind of "saving knowledge" that involves the unconscious as well as the conscious aspects of the human psyche. Just as the material universe had, from its beginning, the naturalistic potential to give rise to humans, so humans may be seen as having the potential to be the receivers of God's soteriologically oriented self-revelation and to respond in an appropriate way.

From the perspective of this kind of model, it is not only the more spectacular kinds of revelatory experience that may be seen in a new light. Membership of the Christian community, for example, is rooted in the way in which each of us who identifies fully with that community has, at some stage of our lives, come inwardly to know the validity of our faith. This "secondary" revelatory experience may have been extremely subtle in its effects, being no more tumultuous, emotionally, than was necessary either to strengthen our existing peripheral membership of the Christian community or to "convert" us to it. This will not usually have involved visions or anything paranormal and may even have occurred over a prolonged period, with none of the emotional elation associated with more rapid conversion. What we have experienced represents, nevertheless, a genuine revelatory action on the part of God, which may be understood in terms of the kind of model I have outlined.

There are, however, clearly aspects of God's self-revelation that cannot be experienced at this relatively low level. In the case of what we may call "primary" revelatory experiences—those that bring into being a new understanding that has not previously existed in a religious community—something more extraordinary is clearly required than the sort of inner affirmation or conversion that the ordinary believer has experienced. For example, the revelation that brought the Christian community into being required (if my analysis is correct) nothing less than a series of complex visionary experiences, which perhaps began during Jesus' lifetime in the experience that we call the transfiguration.

Although the outward circumstances of "primary" and "secondary" revelatory experiences are often very different, they are both understandable in terms of a psychological-referential model, even before that model is expanded in terms of a pansacramental naturalism. But when this expansion is made, they become understandable in a deeper way. From the wider perspective that such an expansion provides, the psychological dimension of both primary and secondary revelatory experiences represents not simply an arbitrarily chosen mechanism that has been used when God 's self-revelation has become appropriate for us. Rather, the psychological potential for revelatory experience can be seen as having been built by God into the potential of the cosmos from its very beginning.

In this naturalistic perspective, a particular revelatory experience need not be seen as the result of divine interference with the natural world. Rather, we may see it as occurring through God's "fixed instructions" for the world. It happens, so to speak, spontaneously: not in the sense that it is arbitrary, but in the sense that it is the "natural" outcome of a particular totality of circumstances. Just as the specific sacraments of the church can be effected only in a particular ecclesial context and the spontaneous emergence and flourishing of a new species requires the context provided by a specific ecological niche, so also, in terms of the kind of pansacramental understanding I have described, we can say by analogy that any particular revelation of God will take place only in what we might call an appropriate "psychocultural niche."

Such a niche has, as we have seen, a cultural component. Only in the context of certain culturally conditioned expectations and needs will any particular revelatory experience arise. Another aspect of the psychocultural niche is a particular sort of psychological openness to God, related intimately to the essentially contemplative dimension of revelatory experience that we have noted in Rahner's understanding. To speak of the spontaneity of any such experience does not, therefore, imply arbitrariness in relation to the cultural situation in which it occurs or to the particular individuals to whom it does occur. Revelation may be spontaneous insofar as it can happen to any person or group of people simply by virtue of the nature of their God-given psyche. Those to whom it happens are still "chosen," however, not in the sense of an isolated act of God's will, but in the sense that its realization requires, of those to whom it occurs, a particular cultural and spiritual readiness.

At one level, we Christians can apply this notion of the psychocultural niche straightforwardly to the historical situation that gave rise to our faith. For according to traditional Christian theology, the new covenant was able to come about not only because of the disciples' direct experience of Jesus, but also because they were able to understand all that they experienced of him "according to the scriptures." Thus, even from the most traditionalist perspective, we can affirm that the psychological aspect of Jesus' direct impact on the disciples was complemented in a complex way by the cultural context provided by their Judaistic background.

However, potential problems are inherent in speaking in this way, for by regarding the Jews as the people of God who were, through the old covenant, being prepared to receive the new covenant, this traditional understanding clearly tends toward a kind of exclusivism. God's self-revelation is seen as being directed toward a particular group of people in the ancient world, and other peoples and cultures can, as a result, easily be seen as irrelevant to God's purposes except insofar as they impinged on this group. This conclusion is clearly in tension with the way in which a pansacramental naturalism assumes that all people have, simply by virtue of being human, the potential to be the receivers of God's self-revelation.

This tension might perhaps be sidestepped in a pansacramental understanding through the observation that the God-given potential of creation is never, in practice, fulfilled in a uniform way. Once a potential quality does begin to be actualized—as it did, for example, with the development of human mental capacities in animal evolution—it can clearly develop so rapidly in one particular group that small initial differences quickly become large ones. This observation seems to allow us to see the development of religious understanding in an analogous way: as something that, once it reaches a particular level of development in some particular group, can rapidly gain so much momentum in that group that the competing understandings of other groups are simply left behind. (For example, the early development of monotheism among the Hebrew peoples might easily be seen in this way.)

However, even if such an understanding contains an important element of truth, we must surely be wary of assuming its validity in any simplistic way. After all, the psychological-referential model of revelatory experience that I have outlined is based on a view of the universality of human spiritual potential, and this view, as we shall see, is paralleled very closely by Eastern patristic perspectives. In this context, it would seem premature to assume that only within one particular religious community have there been authentic revelatory experiences of anything other than the most basic kind.

The alternative to this view is, of course, to acknowledge that at least some of the other faiths of the world have arisen through authentic revelatory experience and have a degree of validity comparable to that of the Christian faith. The problem here is that, in the kind of perspective that has been available hitherto, the incompatibility of the doctrines of these faiths with those of our own is difficult to understand unless we take a nonreferential view of doctrinal language. Here, however, as we shall see in the next chapter, one of the main advantages of the psychological-referential model becomes evident. For when viewed in the light that this model provides, the purely instrumentalist understanding of religious language that has hitherto been necessary to a contemporary pluralism may be jettisoned.

8

The Faiths of the World and the Action of God

For some, the suggestion that God's self-revelation has occurred outside of the Christian faith community, and of the Judaistic one that preceded it, is disturbing. They see such a suggestion as manifesting disregard for the central Christian insight that God's ultimate self-revelation is to be found in a particular historical individual, Jesus of Nazareth. Often, however, this kind of judgement is based on the kind of diluted version of the doctrine of the incarnation that we have noted. It assumes that the *Logos* (Word) of God, of which the Fourth Gospel speaks, is essentially absent from the world except in the person of Jesus. As we have seen, however, the author of the Fourth Gospel had no such understanding, for the *Logos* was understood by him as having been present in the creation from the beginning and as acting throughout human history in the enlightening of every person (John 1:1–9).

Moreover, as we shall see presently, this Johannine understanding of the nature of God's *Logos* was taken up by early Christian theologians in a remarkable way. In particular, as Philip Sherrard has stressed, in important strands of Eastern patristic thought, the incarnation was seen not as "something that occurred only in the unique case of the historical figure of Jesus," but as something that "involves human nature as a whole and so something in which every individual participates."[1] This concept of incarnation, Sherrard insists, is the truly traditional one, and recognition of this, he says, has immediate implications for our understanding of the faiths of the world. The narrow view of the incarnation that has largely prevailed among conservative Christians is, he argues, in conflict with a truly traditionalist view and must "be replaced by a theology that affirms the positive attitude implicit in the writings of Justin Martyr, Clement of Alexandria, Origen, the Cappadocians, St. Maximos the Confessor and many others."[2]

Sherrard's approach here represents an interesting recent phenomenon: the development of a Christian pluralism or inclusivism that is based, not on a facile relativism, but on a central aspect of the Christian tradition itself. A proper understanding of the incarnation, according to this view, implies that the *Logos* "is hidden everywhere, and the types of His reality, whether in the forms of persons or teachings, will not be the same outside the Christian world as they are within it."[3]

We shall return to this view presently, and I shall argue for its essential validity. At this stage of our exploration, however, we need to focus not on the implications of the doctrine of the incarnation for our understanding of other faiths, but on the implications of the psychological-referential model of revelatory experience that I have outlined. This model is, I believe, particularly relevant to the task of developing an adequate theology of the other faiths of the world, which can supplement in an important way considerations based more directly on the Christian tradition.

As we have seen, the pansacramental version of this psychological-referential model, despite its vocabulary of "spontaneous" revelatory experience, cannot be taken to imply that authentic revelatory experience should be seen as arbitrary from a divine perspective. Just as the emergence of a new species requires the particular context provided by a specific ecological niche, so any particular revelation of God will take place only in an appropriate psychocultural niche, which may be defined by the cultural assumptions and the individual psychological makeup of those able to experience some religious revelation or enlightenment. There is a contemplative element in authentic revelatory experience that is linked to these cultural and psychological factors in a way that requires a nuanced and subtle understanding.

We have yet to note an important consideration about this comparison with ecological niches: its implication that particular psychocultural niches not only provide the necessary human environment for some particular revelation to arise, but also limit the type of experience that can arise. Just as a particular ecological niche restricts the kinds of new biological species that can emerge and spread, so, it would seem, a particular psychocultural niche may be seen as having an effect on what kinds of religious experience will be possible.

The analogy here is to the way in which we would expect only certain types of species to have emerged in any particular ecological niche. Just as polar bears could not have emerged as a species other than in a polar region, so, in this view, any particular religious faith will have been able to emerge only in a certain culture, and within that culture, only in certain individuals. Thus, for example, the revelation manifested in the person of Jesus is seen, in traditional Christian theology, as one that required a particular sort of preparation through the old covenant. The advantage of the psychological-referential model of revelatory experience that I have outlined is precisely that it sets this cultural dimension within a wider psychological understanding of revelation, providing a ready way of visualizing why that particular revelation crystallized not only within a particular culture at a particular point in time, but also only occurred among certain individuals in and through a particular sort of experience.

Moreover, because the possibility of religious conversion—the acceptance of a particular revelatory "story"—is clearly linked to the psychological and cultural factors that made that story's initial emergence possible,[4] this model has further explanatory features. In much the same way that it explains how the Christian revelation could have arisen only within Judaism at a certain stage of its development, it also makes understandable the fact that this revelation had its most profound secondary impact in the Hellenistic world and not in other regions to which early Christian missionaries also went. There is a direct analogy with the reasons that a species would be expected to flourish after its emergence only in a certain type of environment. Thus, the development of Hellenistic Christianity may be seen, in "niche" terms, as equivalent to the successful adaptation of a species—sometimes with important modifications—to an environment other than that in which it emerged. The relative failure of Christian evangelism in other areas can similarly be understood in niche terms, as a failure in adaptation—analogous, for example, to the failure of polar bears to flourish in more southerly regions than they do.

The way in which a biological species is no longer to be found in its original geographical location because of ecological changes also has its parallel here. Thus, the dying out of the original Judaistic Christianity may be seen in terms of the way in which the psychocultural niche provided by early-first-century Judaism was so radically changed, both by the emergence of Christianity and then by the effects of the fall of Jerusalem in the year 70, that Judaistic Christianity could no longer flourish and in fact gradually died out.

Thus, the core of the idea of the psychocultural niche is that, just as life is potentially multiform and will arise and develop new forms spontaneously through natural (chemical and biological) processes, in accordance with the possibilities inherent in a given ecological environment, so God's self-revelation also is potentially multiform. It too will arise and develop new forms spontaneously, through natural (psychological) processes, in accordance with the possibilities inherent in a given cultural environment. Whether we are considering life or revelation, however, neither spontaneity nor naturalness precludes a theological explanation in terms of divine action through the sacramental potential of the cosmos. Rather, as we have seen, both life and revelation may be held to represent new irreducible emergents intrinsic to that sacramental potential. As such they remain, in the deepest sense, gifts of God.

However, we also have now seen that revelation may be a far more complex gift than has hitherto been generally appreciated. In particular, the question of which aspects of any revelatory experience are genuinely referential, and which are simply part of the psychological vehicle by which they have been appropriated, cannot be answered easily.

As we have noted, for example, it becomes at least conceivable that the referential content of the Easter experiences of Jesus' disciples may not only exclude the concept of Christ's bodily ascension into heaven, but even of his "resurrection body" as such. Indeed, it would not be illogical, within the general framework of the model, to argue that it might exclude any information at all about his person. (The genuine referential content of those experiences might, for example, be related to something much more abstract, such as the nature of death and eternal life or of the incarnational relationship between God and humanity.) In this view, at least some of what has traditionally been "read off" from the experiences of the risen Christ may represent not referential truth, but simply evocative mythological categories that are potent in bringing to salvation those who respond to them in faith.

If this were the case, however, then it is clear that the genuine reference made available to Christians through the Easter appearances might have been made available to others—partially or wholly—through a completely different revelatory vehicle, with no reference to the historical Jesus at all. If the reference in the resurrection appearance experiences were indeed to something rather abstract, such as the nature of death and eternal life, then many religions have at their heart some revelatory or enlightening experience related to these or similar concepts.

This sort of consideration makes clear how apposite a two-component model of revelatory experience is to interfaith dialogue. It allows us to see clearly how the revelatory or enlightening experiences from which all the main faiths of the world have emerged are in principle susceptible to analysis in terms of the model. Moreover, questions about the referential content of such experiences, as they arise for each faith from such a model, clearly impinge directly on the questions about the relationship between faiths that are at the heart of interfaith dialogue. Indeed, the three main existing approaches to the nature of doctrine, which already inform attitudes to that dialogue, may now be examined from the perspective of a wider framework than hitherto.

The first of these existing approaches is the exclusivist one, in which all religious traditions but one's own are viewed as having arisen from processes other than God's genuine self-revelation. Within a psychological-referential framework, however, this approach would be justifiable only if all revelatory experiences except those at the roots of (or conforming to) one's own faith could be assumed to be "merely" psychological, at best effective at an instrumental level. In this view, only the experiences associated with one's own tradition should be seen as referential to an extra-psychic reality.

The second existing approach to interfaith dialogue is the complete opposite of this exclusivist one. It is the relativist one of viewing all faiths as essentially equal in terms of their referential validity. A psychological-revelatory version of such a view would presumably hold either that no faith has arisen from genuinely referential revelation, or else—from the point of view of what has been called a theocentric relativism[5]—that all revelatory experiences have acted equally as signposts to the transcendent dimension of existence but have no further referential content. For both of these views, the specific differentiating content of the faiths that have emerged from primary revelatory experiences would be interpreted in purely instrumentalist terms and would not be susceptible to comparison in terms of the question "Which is true (or at least more true)?"

From the point of view of a psychological-referential model of revelatory experience, however, both the relativist approach and the exclusivist one manifest an essential arbitrariness. While the exclusivist view, on purely subjective grounds, limits referential content to certain revelatory experiences, the relativist one is equally arbitrary in its refusal to consider any claim to specific referential content.

This refusal has been encouraged in recent years by what is sometimes called a "linguistic" approach to religious life and doctrine, such as that advocated by George Lindbeck.[6] This approach, with its stress on the way in which doctrine, religious narrative, ritual, and life orientation are mutually interpenetrating, stresses that religious faith is to be understood primarily in relation to the believing community as manifested in all its activities. Such a stress undoubtedly has, in the context of a psychological-referential model, important points, not least in relation to its potential clarification of the cultural dimension of that model. However, the linguistic view's instrumentalist understanding of doctrine is such that the questions about reference in revelatory experience, which we are addressing here, are bypassed in an essentially arbitrary manner. What is surely required (and indeed may be legitimate even in terms of its Wittgensteinian background[7]) is an expansion of the linguistic view in terms of the questions about reference that arise from a psychological-referential model.

Such an expansion would certainly be possible and indeed seems to be at least implicit in the third sort of approach to interfaith dialogue that we need to consider. This is the understanding that, while holding that religious languages may make referential truth claims that are explorable, at the same time recognizes the complexity of the way in which these truth claims are embedded in particular cultures. In this view, the existence of competing religious frameworks necessarily means that some religious traditions may possess certain truths (or at least approximations to truth) more fully than do others, but it equally means that no one faith can make exclusivist claims that it has nothing to learn from the others.

Certainly, only an approach of this sort can avoid the essential arbitrariness of the exclusivist and relativist approaches. However, it has hitherto suffered from a lack of focus on the revelatory or enlightening experiences on which the various faiths of the world are based. This problem finds a ready resolution in the context of the psychological-referential model that we are exploring here, so it can be argued that an essentially new version of the pluralist position becomes possible. For such a model suggests that referential truths about some aspect of the divine reality may have been vouchsafed in revelatory experience at different times and in different cultures, in such a way that the major faiths of the world emerged. The differences among those experiences, as we have seen, may be viewed as having been largely determined by the differences among the cultures concerned, with their very different expectations about the nature of salvation and of the ways in which it might be attained or given. Thus, the differences among the world's faiths are, for this model, a natural corollary of the psychological mode by which those faiths have been engendered.

In practice, a spectrum of views can be described in these terms. At one end of this spectrum is the sort of "convergent pluralism" that begins, methodologically, from an essentially relativist position but does not assume that this position must necessarily be maintained as dialogue proceeds. At the other end is the sort of "inclusivism" that takes as its starting point the belief that the other faiths of the world may implicitly manifest a hidden and incomplete version of one's own faith. Thus, for example, one of the best-known Christian studies written from this viewpoint is entitled "The Unknown Christ of Hinduism."[8]

The problem with this latter starting point is, of course, that it can easily manifest itself as a form of condescension to those of other faiths, which is not only patronizing but also at least partially blind to what can be learned from them. Even when we recognize this problem, however, it is difficult to see how any other fruitful starting point can be adopted. For we cannot, in practice, come to interfaith dialogue without convictions rooted in the particular faith community to which we belong. If we did, then we would not be entering a real dialogue with others at all, but merely using them to provide information about their own faiths in order to undertake an abstract philosophical investigation from what we believed to be a neutral position. As we have seen in our investigation of natural theology, however, a truly neutral starting point can lead us nowhere, even in relation to the investigation of a single faith. If theology is "faith in search of understanding," then it can only start from a real faith that is expressed, however provisionally, in the form of certain convictions.

Indeed, it can be argued that we must enter interfaith dialogue not only with convictions rooted in our own faith community, but also with a conservative reluctance to change these convictions without very good reason. As I have argued elsewhere, there are good philosophical grounds to adopt this as a methodology, since parallels between scientific and theological language usage suggest that the scientists' reluctance to abandon well-established theory may be taken as normative for the development of theological doctrine as well.[9]

A conservative reluctance to change without good reason is, however, very different from the sort of fearful conviction that we can learn nothing about certain of our doctrines from those of other faiths. If we adopt this latter position, we will inevitably reach conclusions that merely echo our initial assumptions. If, in contrast, we enter dialogue with real humility—with a commitment to that element of a "convergent pluralism" that recognizes that "most, and probably all, traditions" are in need of revision if they are "to approximate more nearly to the fuller unitary truth which none of them yet fully encapsulates"[10]—then dialogue will have a chance of bearing real fruit. For we shall, if we take this position, be open to seeing our own Christian faith in a broader perspective and will recognize that such a broader perspective may be necessary for legitimately seeing other faiths as being "included" in it. In this kind of "neo-inclusivism" (as we might call it), our approach will not be one of attempting to fit other religions into our own as we at present understand it, but rather of recognizing that we may not yet understand our own faith well enough to make such an attempt.

This neo-inclusivist entry into dialogue with others will not, however, simply be entry into an intellectual process, for one of the most important things about interfaith dialogue is its experiential effect on those who participate in it. (Indeed, this effect is not limited to formal dialogue, for anyone in a pluralistic society can experience it through simple social encounter.) If we meet the best representatives of other faiths in the spirit of humility I have described, we can only, in my view, come to feel a new conviction of the truth of the Fourth Gospel's notion that the *Logos* "enlightens everyone" (John 1:9) and a new appreciation of the patristic expansion of this notion, which we have noted in the context of Philip Sherrard's views.

At the same time as we examine the general implications of this conviction, however, we need to ask ourselves more specifically what it might mean to accept Sherrard's view that the *Logos* "is hidden everywhere, and the types of His reality, whether in the forms of persons or teachings, will not be the same outside the Christian world as they are within it."[11] In particular, our understanding of other faiths of the world will inevitably depend on how we understand the nature and role of their and our doctrinal languages. In our exploration so far, this question has been approached largely from an apophatic perspective; I have emphasized what our doctrinal language is *not* intended to be. What we must now explore is the question of how we are to develop a corresponding "cataphatic" element of our understanding. We must ask what the positive role of our religious language is, especially when viewed in the context of the other faiths of the world.[12] The key to seeing this positive role lies, I suggest, precisely in the details of the kind of approach that Sherrard has developed. Therefore, it is to these that we must now turn.

9

The Faiths of the World and the *Logos* of God

For most people in the Western Christian world, the name of Philip Sherrard is familiar, if at all, only because of his nontheological writings, his pioneering exploration of a theology of human sexuality, and his somewhat eccentric views about the effects of scientific thinking. In the Eastern Orthodox community to which he belonged for most of his adult life, however, Sherrard's reputation rests on a much wider range of writings and translations. As one of his obituarists put it in 1995, he may be seen as standing, in relation to Orthodox spirituality, "in the vanguard of a revival in both East and West which 20 or 30 years ago would have seemed scarcely imaginable."[1]

The reaction to Sherrard among his fellow Orthodox has, however, often been somewhat ambiguous. They usually recognize that he is, as Bishop Kallistos Ware put it in the foreword to Sherrard's last (and posthumously published) book, "a creative and sometimes prophetic interpreter of the living tradition of the Orthodox Church."[2] In certain respects, however, at least some of them feel uncomfortable with aspects of Sherrard's thought, and this discomfort is shared by many of the conservative Christians of the West who know his work.

One of the main reasons for this discomfort is indicated by the way in which, even in his celebratory foreword, Bishop Kallistos clearly feels the need to mount a defense of his old friend. The problem is, as the bishop notes, that some who have encountered Sherrard only in an explicitly Christian context "have been surprised and even disturbed by his openness to non-Christian traditions." In particular, he goes on, when reading Sherrard's posthumous book, *Christianity: Lineaments of a Sacred Tradition*, they may well ask why he does not speak "more clearly about the New Testament, the Church and the sacraments, about Christ's Incarnation, Crucifixion and Resurrection." Indeed, the bishop notes, they may ask, "How far does he uphold the uniqueness of Jesus Christ as the *only* incarnate Son of God?"[3]

For Bishop Kallistos, Sherrard's approach in this posthumous book is defensible provided that we keep two things in mind. First, he says, we must remember that much of Sherrard's writing was "addressed in the first instance . . . to a mixed audience that included adherents of different faiths, as well as 'seekers' who had not yet found a home in any particular tradition." A terminology that was exclusively and aggressively Christian would, he suggests, have risked alienating many who might otherwise have been prepared to listen attentively to Sherrard's message. Second, "and more importantly," he says:

> Philip's writings form a unity, and each particular chapter or book has to be read in the context of his total *oeuvre*. As soon as this is done, it becomes overwhelmingly clear that he does not regard Christ's Incarnation as secondary or peripheral. On the contrary, his entire understanding of the relationship between the uncreated and the created—between God, humankind and the world—is based upon one foundation or paradigm, and upon one alone: upon the "union without confusion" between divine and human nature that came to pass in the single, undivided person of Christ incarnate.[4]

For some, however, this defense of the consonance of Sherrard's last book with a traditionalist Christianity will not be entirely convincing. While it is true, for example, that several of its chapters began their life as texts written with a mixed audience in mind, it is also true (as Bishop Kallistos notes elsewhere in his foreword) that Sherrard not only gathered together articles and lectures composed over a period of many years as the basis for the book, but also "revised the whole text, making several significant additions." Because of this, as the bishop himself acknowledges, "Although the book cannot be termed in the strict sense [Sherrard's] 'last will and testament,' it reflects nevertheless his considered viewpoint at the conclusion of his life."[5]

In addition, while the bishop is right in his claim that the Chalcedonian definition was central to Sherrard's thinking, his *apologia* does not note the extent to which Sherrard, when he speaks about the "union without confusion," emphasizes its suprahistorical aspect so strongly that the question of how far he upholds the uniqueness of the historical Jesus is not answered unambiguously. Sherrard may well, as his *Times* obituarist put it, have "steered a scrupulous course between the Scylla of East European 'New Orthodoxy', with its sometimes jingoistic overlay, and the Charybdis of New Ageism, to which, in a different idiom, his ideas at times came close."[6] Whether he was always as successful in keeping to this course as Bishop Kallistos and other traditionalist Christians might have wished remains, however, an interesting question.

The main chapter that deals with these issues in Sherrard's posthumous book is the third, entitled "Christianity and Other Sacred Traditions." This begins with a reiteration of a remark made in the previous chapter: "One of the conditions for any renewal of the Christian Church is that the Church renounces the claim that the Christian revelation constitutes the sole and exclusive revelation of the universal truth."[7] In this previous chapter, this plea is related primarily to the need for the church to overcome its historically conditioned tendency toward what Sherrard calls "coercive authoritarianism."[8] In the new chapter, however, the analysis is not historical and sociological, but theological. It is based on the positive attitude implicit in Paul's speech to the Athenians (Acts 17:22ff.) and on the explicit way in which many Christians of the early centuries (such as Justin Martyr and Clement of Alexandria) saw the *Logos* doctrine outlined in the prologue of the Fourth Gospel as providing the essential foundation for a Christian assessment of non-Christian religion.

In this patristic strand of thinking, Sherrard notes, Greek—and by implication all—philosophies are seen as "fragments of a single whole which is the Logos."[9] This attitude, he observes,[10] is summarized in comments by Irenaeus, for whom "there is only one God, who from the beginning to end, through various economies, comes to the help of mankind,"[11] and by Origen, for whom there is "a coming of Christ before his corporeal coming, and this is his spiritual coming for those men who had attained a certain level of perfection, for whom the whole plenitude of the times was already present."[12]

In this strand of patristic understanding, Sherrard sees an affirmation of what he calls "the universality of the incarnation."[13] For, he says, what complements and underpins "this doctrine that prior to his advent in the flesh the Logos is present in man and in all creation, and may be apprehended by all men irrespective of time and place, is an understanding . . . [which] affirms that through the Incarnation the divine Logos incorporates Himself not in the body of a single human being alone but in the totality of human nature, in mankind as a whole, in creation as a whole."[14] This strand of thinking has become unfamiliar to most of us, he believes, largely because it has been eclipsed by "a linear view of history . . . which sees Christianity as a series or succession of salvation events destined to culminate in the appearance of Christ." This latter view, he says, "tacitly ignores the idea of an ever-present eternity that transcends history."[15]

For Sherrard, an emphasis on historical salvation events as manifesting an eternal reality is crucial: "If Christians are to integrate other religions, positively and creatively, with their own doctrinal perspective, they have to go beyond and discard the concept of linear 'salvation history.' " This concept must, he insists, "be replaced by a theology that affirms the positive attitude implicit in the writings of Justin Martyr, Clement of Alexandria, the Cappadocians, St. Maximos the Confessor, and many others. The economy of the divine Logos cannot be reduced to His manifestation in the figure of the historical Jesus."[16] What is required, he goes on, is a recognition of the way in which "the Logos in His *kenosis*, His self-emptying, is hidden everywhere, and the types of his reality, whether in the forms of persons or teachings, will not be the same outside the Christian world as they are within it. Yet these types are equally authentic: any deep reading of another religion is a reading of the Logos, the Christ. It is the Logos who is received in the spiritual illumination of a Brahmin, a Buddhist, or a Moslem."[17]

The fact that these people of different religious traditions have not received precisely the same form of illumination as the Christian is, for Sherrard, readily explicable. He says it is necessary for the Truth at the heart of all things to express itself "in a way which is accessible to the human intelligence. It has to take account not only of the human state as such, but also of the various, though relative, divisions within mankind." Put in its broadest terms, he goes on, "one might say that the differences between the various traditions are due to the differences in the cultural *milieux* for which each is providentially intended and to which each is therefore adapted."[18]

Certainly, he says, we must not abandon the concept of "an underlying metaphysical order, a series of changeless and universal principles from which all derives and on which all depends." However, he goes on, we must recognize that "because of the inherent limitations of the human state, we cannot recognize and realize [these principles] directly, in their naked essence. We can only be gradually, and stage by stage, initiated into them." Once this is recognized, he says, we will be able to see how, in various times and places,

> the Supreme, either through direct revelation of a Messenger or Avatar, or through the inspired activity of sages and prophets, has condescended to clothe the naked essence of these principles in exterior forms, doctrinal and ritual, in which they may be grasped by us and through which we can gradually be led into a plenary awareness of their preformal reality. These forms may be many . . . but beneath this multiformity may always be discerned, by those who have eyes to see, the essential unity of the unchanging, non-manifest, and timeless principles themselves.[19]

Sherrard denies that the common factor he perceives in all authentic religious traditions is susceptible to an analysis in which the doctrines of the different faiths can be reduced to some underlying and more fundamental doctrine. What is common to the faiths of the world can be perceived only through an essentially spiritual perception rooted in a particular tradition. Thus, religious believers can perceive what their own faith has in common with other faiths primarily when they approach the question as contemplative adherents of a particular sacred tradition, not as intellectually acute philosophers.

At first sight, Sherrard's way of expressing this common factor—in terms of a variety of manifestations of the incarnate *Logos*—may give the impression that he is advocating nothing more than the usual sort of inclusivism, in which other religions are seen merely as hidden and incomplete versions of Christianity. His view is, however, rather more subtle than this first impression may suggest, for he holds that we simply cannot answer the question of which religion is superior to all the others, since this would require the possession of a kind of knowledge that divine revelation does not, and cannot, give us.

This is the case, Sherrard insists, for two reasons. First, he says, God must be seen as transcending "all forms, whether intelligible, imaginable or sensible. In this sense He is beyond all determination and limitation. But if He is to reveal Himself to human beings in their fallen state, He has to determine Himself, and hence to limit Himself, in a specific intelligible, imaginable or sensible form." Because of this self-limitation, "however many the forms in which [God] reveals Himself, aspects of His full reality will always remain undisclosed."[20] Second, he goes on, we must recognize the limitations of our mode of consciousness. Humankind is the "locus or medium" of God's self-revelation, "and a corollary of the affirmation that man is created in the image of God is that each single individual, let alone the cultural and linguistic group to which he belongs, will express this image, and therefore the divine qualities that constitute it, in a unique way." Because of this, he says, each of us must recognize that "God is not limited to the mode in which He is epiphanized in me or to the form in which I am capable of perceiving Him."[21] For others, this mode and form may be very different.

This kind of reasoning leads Sherrard ineluctably to a position that is not inclusivism of the usual kind. Rather, what emerges from Sherrard's thinking is a position that, at least implicitly, cannot exclude the possibility of faiths other than Christianity being fuller expressions of the divine reality. Moreover, this implicit suggestion is reinforced by some of Sherrard's specific comments on the Christian revelation. For, he says, the way in which God has been revealed in the historical Jesus is related intimately to the "state of blindness, ignorance and ineptitude" of the particular section of humanity to which the earliest Christians belonged, and to which most Christians still belong.

Both the incarnation and the crucifixion are, he says, "the consequences of our ignominy." Had the type of consciousness that typified the first-century Judeo-Hellenistic world "been capable of receiving that revelation in a more subtle and more spiritual form, then it would have been communicated in such a form."[22] Thus, Sherrard goes on:

> The fact that the revelations of other sacred traditions are not centred in and do not depend upon an incarnation equivalent to that of Christ does not mean that the wisdom enshrined in their doctrine is spurious or false, or is in any way inferior to the wisdom enshrined in Christian doctrine. It may simply mean that the consciousness of the *milieux*, human and cultural, to which these revelations are given is of such a type or quality that God does not have to manifest Himself in a visible, historical form in order to communicate a true knowledge and understanding of things.[23]

The last part of this statement may take us to the heart of Sherrard's thinking. In his reflection about divine revelation, including that manifested in the person of the historical Jesus, he focuses entirely on the way in which it leads to true spiritual knowledge: to *gnosis*. It is eloquent of his whole approach, in fact, that the index to his last book contains more references to this term than to any other except the related entry "God, Manifestation of image or nature, to man."

As the first chapter of this posthumous book makes clear, an authentic sacred tradition is, for Sherrard, to be recognized as such primarily by its manifestation of two aspects. The first, he says, "is what we may call the gnostic aspect, in that it pertains to the knowledge of what constitutes the fully realized and perfected human state, of the relationship with the divine that such a state presupposes, and of the corresponding relationship between God, ourselves and the natural world." The second aspect is that which "may be called the aspect of spiritual practice, or spiritual method, through which we can attain the state of contemplation and glorification in which the purpose of sacred tradition is consummated."[24]

This gnostic emphasis does not, we should note, mean that Sherrard's approach should be equated with that of the heretical Gnostics of the early Christian centuries, whose approach presupposed the existence of "secret" doctrines known only to an initiated elite. It can be seen, rather, as comparable to the understanding of the Christians of those centuries who, particularly in the Alexandrian tradition, stressed the importance of a genuinely Christian gnosis: one depending not on secret traditions, but rather on an ever-deeper apprehension of the spiritual realities underlying the public proclamations of the church.

Even if we agree with Sherrard in recognizing the validity of this kind of Christian gnosticism, however, we must still ask why he puts little stress on certain aspects of the church's public proclamation—especially all that pertains to the life, death, and resurrection of Christ. Moreover, the questions that arise from this apparent lacuna in his characteristic approach are underlined by some of his specific (and perhaps more eccentric) comments. At times, for example, he seems to regard Jesus' crucifixion as something to be understood not, as traditional Christianity would insist, in relation to a fallen humanity as such, but only in relation to a particular segment of it. Indeed, besides aguing—as we have noted previously—that God may not have needed to manifest God's self to all humanity as in the Judeo-Hellenistic world, he also suggests quite explicitly that if God had "manifested himself in such a form in these other *milieux*, He would not have been crucified."[25]

If we are to understand this aspect of Sherrard's thinking, what is important to recognize is that the crucifixion has not, for him, ceased to be an essential aspect of the Christian proclamation. He seems to suggest, however, that its primary purpose within that proclamation is to serve as a focus for contemplation, through which the mature Christian will eventually come to a full recognition of a universal and timeless reality. This reality is, for Sherrard, God's self-emptying, God's *kenosis*, which he assumes will be manifested in some way in all genuine religious experience. Indeed, he explicitly states that a focus on divine *kenosis* is a characteristic not only of Christian contemplation, but also of "that of Hinduism, Buddhism, and Islam."[26]

This approach ties in very clearly with Sherrard's view that the salvation history of any particular tradition, including Christianity, is essentially a set of manifestations of an ever-present and timeless reality. This is particularly the case, in fact, when it comes to his way of speaking about the incarnate *Logos*. For as we have already seen, Sherrard's understanding of this term is by no means limited to its application to the historical Jesus, and what this means for him is underlined when, at one point in his essay, he goes as far as to say, "The idea of God-manhood possesses a significance that is intrinsic to human nature as such, *quite apart from its manifestation in a historical figure who exemplifies it.*"[27]

At least implicitly, it would seem, Sherrard is leading us to the conclusion that the "union without confusion" at the heart of his thinking is not *necessarily* manifested in a unique and definitive way in the incarnation in Jesus. That union is, for him, first and foremost an eternal reality, of which Jesus is the prime manifestation for Christians. Whether Jesus is the only manifestation in this incarnate form is, however, left as an open question, and it is not clear whether Sherrard is consciously or unconsciously hinting at his own answer to it when he speaks of manifestations of the *Logos* "in the forms of persons or teachings"[28] and mentions in passing the Hindu concept of an "Avatar."[29]

10

Art and Sacrament

Much that I have said in the last few chapters can be summed up in some words of Philip Sherrard quoted in the last chapter, which spoke of the way in which the divine *Logos* is hidden everywhere and of how the types of the reality of this *Logos*, whether in the forms of persons or teachings, will not be the same outside the Christian world as they are within it. Sherrard goes on to say that not only do we need to affirm that the types of God's reality, as perceived by other faiths, may be as "authentic" as our own, so that any deep reading of another religion may truly be "a reading of the Logos, the Christ," but we also need to recognize that God has "at different times and in different places . . . condescended to clothe the naked essence of [the principles underlying all reality] in exterior forms, doctrinal and ritual, in which they can be grasped by us and through which we can be led into a plenary awareness of their preformal reality."[1]

This distinction between "naked essence" and "exterior forms" is, in fact, crucial to the approach I advocate. For it reflects my appropriation of the Eastern apophatic tradition and also of Yves Congar's stress on the way in which God has been made known to Christians, not in abstract knowledge, but in "signs" that are always oriented primarily toward salvation. As we have seen, these signs have a genuine ontological content, according to Congar, but this can only be expressed in terms of what he calls "mysteries," which are partially hidden truths. These are made present to us most fully not in linguistic statements (which must always be understood apophatically), but in liturgical celebration.

Thus, in this approach, what reflects our knowledge of God is not primarily our abstract theological thinking, but our spirituality in all its manifestations. In particular, as emphasized by both the Eastern Orthodox tradition that Sherrard reflects and the Roman Catholic one that Congar expounds, the common prayer of the people of God, by informing and encompassing all private prayer, is the prime reflection of our response to God's self-revelation in Christ.

Central to this common prayer is, in these traditions, the celebration of the sacraments—or as the Orthodox tend to call them, the "mysteries." Here, however, we now face a problem. While the importance of the sacraments is still maintained in most liturgically oriented traditions, our forebears' instinctive sense of their centrality has now begun to evaporate. Although we may still be able to parrot the various catechetical formulas about sacraments being "an outward and visible sign of an inward and spiritual grace" or "effecting what they signify and signifying what they effect," a quasi-instinctive sense of the intrinsic connectedness of signification and effect is no longer common among us. All too easily, this linkage is seen, not as an intrinsic aspect of our created being, but as having been imposed by God, somewhat arbitrarily, as a sort of didactic aid. We have largely lost what Andrew Louth calls a sense of "the architechtonic significance of the image in the created order."[2]

What Louth means by this term is conveyed in his description of the way in which, for the patristic tradition—and especially for John of Damascus—images and signs of various kinds reflect the "multitude of ways in which reality echoes reality." In John's thinking, he says, images "establish relationships between realities within the Trinity, between God and the providential ordering of the universe, between God and the inner reality of the human soul, between visible and invisible, between the past and the future, and the present and the past," so that "the image, in its different forms, is always mediating, always holding in harmony."[3]

Nowadays, however, few of us can respond positively to this way of conceiving reality, and our inability to think instinctively in this way remains even when we recognize, with Louth, that for John this understanding is "both undergirded by, and finds its highest expression in, the truth of the Incarnation." It is observable that even those who have a relatively sophisticated understanding of the incarnation still, in our own time, find it difficult to understand the notion that, when God became incarnate and "even more" (as Louth puts it) "when at the Last Supper he gave his disciples bread and wine as his Body and Blood, God 'placed himself in the order of signs'. "[4]

There is, however, at least one way in which we can begin to explore this notion: through the thinking of a person who, in the twentieth century, was significantly influenced by Maurice de la Taille's concept of the "order of signs," which Louth invokes here. This person is the artist and poet David Jones (1895–1974), whose feeling for the world of signs Louth sees as reflecting, at least in some degree, that of John himself.[5]

One of the most significant events of Jones's life was his first glimpse, after a Protestant upbringing, of a Roman Catholic Mass. He was, at the time, a private soldier in the Royal Welch Fusiliers, serving near the front line during the First World War. Foraging for firewood, he came across a ruined outhouse and, peering through a chink in the wall, he saw (as he put it in a letter written near the end of his life) "not the dim emptiness I had expected but the back of a sacerdos in a gilt-hued *planeta* . . . two points of flickering light . . . white altar cloths and the white linen of the celebrant's alb and amice and maniple." He went on to explain how the scene had made "a big impression" on him:

> For one thing I was astonished how close to the Front Line the priest had decided to make the Oblation and I was also impressed to see Old Sweat Mulligan, a somewhat fearsome figure, a real pugilistic, hard-drinking Goidelic Celt, kneeling there in the smoky candlelight. And one strong impression I had . . . I felt immediately the oneness between the Offerant and those toughs that clustered round him in the dim-lit byre—a thing I had never

felt remotely as a Protestant at the Office of Holy Communion in spite of the insistence of Protestant theology on the "priesthood of the laity."[6]

Jones went on to become a Roman Catholic, as well as a sort of amateur theologian, whose letters to the press, essays, and reviews often contained a theological theme. When some of these were collected together in a volume entitled *Epoch and Artist,*[7] its editor, Harman Grisewood, chose to put on the title page an unattributed quotation: "He placed himself in the order of signs." It was an entirely appropriate quotation, for this was precisely what Jones had done throughout his adult life, both as artist-poet and as Christian. In Jones's view, it was the sign-making nature of the human condition that made possible both human creativity and the sacramental understanding that was central to his faith.

The quotation chosen by Grisewood, however, did not originally refer to any artist or poet in the usual sense. In fact, as we have noted, it came from the work of the theologian Maurice de la Taille, and it referred to Christ himself.[8] What de la Taille had meant when he talked about "the order of signs," in relation to the intrinsic link between the Last Supper, the cross, and the anamnesis of the Eucharist, became a central aspect of Jones's understanding. For, as Jones noted in his essay "Art and Sacrament," de la Taille's thinking had "shed a sort of reflected radiance on the sign world in general."[9]

Jones felt, however, that on its own, this theological perspective was insufficient. It required expansion, he believed, in terms of what seemed to him a prior question that was "anthropological rather than . . . theological": that of why "*men make sacraments.*"[10] The answer, he went on to suggest, lies in the fact that signs, rites, commemorative acts, and the like are used by man, not only because it is "natural to him," but more specifically—and this is central to his whole thesis—because it is "natural to him by virtue of his being an artist."[11]

Jones then went on to recall how, soon after leaving the army, he resumed the formal training as an artist that he had begun before the war and pondered questions about art and about the Eucharist. "The question of analogy" he recalled, "seemed not to occur until certain Post-Impressionist theories began to bulk larger in our student conversation. Then, with relative suddenness, the analogy between what we call 'the Arts' and the things that Christians called the eucharistic signs became . . . apparent. It became increasingly apparent that this analogy applied to the whole gamut of 'making'. "[12] In particular, Jones went on:

> [Something implicit in the Postimpressionist theories then current] opened the eyes of us to what, many years back, I had the occasion to describe as "the unity of all made things." For one of the more rewarding notions implicit in the post-Impressionist idea was that a work [of art] is a "thing" and not (necessarily) the impression of some other thing. For example, that it is the "abstract" quality in a painting (no matter how "realistic") that causes that painting to have "being," and which alone gives it the right to be claimed an art-work, as a making, as *poiesis.*[13]

On its own, however, this insight was not sufficient for Jones. Acknowledging that "the post-Impressionist theories indicated an approach that was most salutary," he complained that they "also provided ammunition for an unrewarding and somewhat unreal battle . . . the war of theories concerning 'abstract art' and 'representational art'. " It is necessary, he went on, to assert "as axiomatic that all art is 'abstract' and all art 're-presents'. "[14] The hyphen in *re-presents* was, for Jones, crucial. Although the "reality" conveyed in any work of art may be complex, he said, the work itself is "a 'thing,' an object contrived of various materials and so ordered . . . as to show forth, recall and re-present [that reality], strictly within the conditions of a given art and under another mode. . . . It is a *signum* of that reality and makes a kind of anamnesis of that reality." Thus, for Jones, if the anamnesis of the Eucharist is instrumental in making Christ really present in the sacrament of the Eucharist, the reality conveyed by a work of art becomes truly present to the beholder—if not "in the particular sense used by the theologians," then at least "in a certain analogous sense."[15]

For Jones, this understanding should apply to more than "works of art" in the narrow sense. He had a belief very similar to that manifested in Ananda Coomaraswamy's dictum that an artist is not a special kind of man, but every man is a special kind of artist. (Jones had no doubt heard this saying often from the lips of Eric Gill, who was a strong early influence on him.) For Jones, things as diverse as "strategy, a birthday cake, a religious rite and a well known picture" all bore witness "to the nature of the thing we call art and the nature of the creature we call man and the inseparability of the one from the other." The activity of art, he went on, "far from being a branch activity, is truncal, and . . . the tree of man, root, bole, branches and foliage, is involved, of its nature, in that activity."[16] Only in this context, he felt, was it possible to make sense of the theological notion of sacrament, which would be "devoid of meaning unless the nature of man is sacramental." Indeed, he asserted, "without *ars* there is no possibility of *sacramentum*."[17]

Jones stressed that the essay in which he made these comments arose primarily from his concerns as an artist and poet and that it could, as a result, constitute "nothing . . . beyond an enquiry."[18] This is a statement that the student of liturgical and sacramental theology will inevitably endorse. Nevertheless, for all its limitations, Jones's thinking in this area still constitutes an important inquiry, and not only because it prompts in us, in A. M. Allchin's words, an awareness of how "we fail to appreciate the nature of the specific sacraments of the Church . . . because we fail to appreciate the way in which all things are sacramental, in particular all man's acts of making."[19] Over and above this prompting, it is arguable that Jones's fundamental insight—that we are essentially sign-making beings—poses for us a number of specific questions that have remained without an adequate resolution.

These can be illustrated, and their central point illuminated, by a question that has troubled many Western Christians: whether some of the liturgical reforms that occurred in Jones's own Roman Catholic Church toward the end of his life (the majority of which he reacted against strongly) were as appropriate as was believed by those who imposed them. I refer here not so much to the change to the vernacular, against which Jones fulminated publicly and articulately. (In his own writing, there had already occurred the problem of conveying historical context in translation, which he saw as a central issue in liturgical translation.[20]) I refer, rather, to the kinds of minor reform, copied in other churches, about which he fulminated privately and less articulately: those involving either a significant change of symbolic emphasis—as in the change of the orientation of the priest during the Mass—or else a reduction of emphasis on "secondary" symbolism.

Perhaps because Jones's published comments on these issues are mostly taken from private letters rather than from documents intended for public debate, he can often, it must be admitted, sound like little more than a saloon bar reactionary when he refers to them. In the letter with which we began, for example, in which he recounted his first glimpse of the Mass, his mention of the maniple led him to fulminate about the way in which that item of the traditional eucharistic vestments had "been abandoned, without a word of explanation, by these blasted reformers."[21] We should, however, resist the temptation to dismiss such comments simply because of their tone. Jones was not simply a splenetic reactionary, for he could see clearly that some of those who shared his views were "rather like those cavalry officers of the 1st World War who were totally blind to the requirements of trench warfare."[22] Rather, the tone he adopted in these letters seems to have been both a characteristic of his general style when writing to intimate friends and a reflection of his genuine bemusement at the reformers' lack of insight into what was, for him, central to all liturgical action: the fact (as he saw it) that humans are, first and foremost, artists, makers, and users of signs. He does seem to have genuinely felt, as he wrote in another letter, that "these blasted liturgists have a positive genius for knocking out *poiesis*."[23]

The question that faces us is whether he may have been right, if only instinctively. Take, for example, one of the issues to which Jones referred explicitly in the letter with which we began our exploration of his views, in which he commented on "Old Sweat Mulligan" and the "oneness between the Offerant and those toughs that clustered round him." The letter continues with an implied criticism of those who "declare that the turning round of the *mensa* . . . made the faithful more at one with the minister and so get back nearer to the Coena Domini [Lord's Supper]."[24] For Jones himself, the matter could be taken no further. For us, however, it is at least possible to argue that the change of orientation induced a real loss.

The old position, with the priest aligned with the people, surely symbolized rather wonderfully the way in which the sacrifice of thanks and praise of the people is united with the total sacrifice of Christ (of whom the priest is here an icon) and offered to God the Father. The classical liturgical pattern[25] embodied in the words of all recent Western liturgies—"To the Father, through the Son, and in the Holy Spirit"—is clearly affirmed in the symbolism of the orientation. This pattern is, moreover, conveyed instinctively to the likes of "Old Sweat Mulligan" as effectively as to those who are able to appreciate it at an intellectual level. The new orientation, by contrast, seems in danger of symbolizing, not the patristic and modern attitude to the nature of the Eucharist, but rather that medieval one in which the classic Trinitarian pattern was obscured through an emphasis on prayer to Christ himself.

It would seem that, just as what Thomas Torrance has characterized as liturgical Apollinarianism[26] has disappeared at the verbal level, it has reappeared at the level of orientational symbolism.

This particular issue is, of course, debatable; not only might aspects of the new orientation outweigh these disadvantages, but "best practice" might indicate that the current orientation of the celebrant need not obscure the classical Trinitarian pattern.[27] All the same, the issue seems eloquent of an attitude that has become almost universal among Western liturgists and was, indeed, explicitly used to justify many of the liturgical reforms of the twentieth century. This is the belief that everything except the *primary* symbolism of any particular liturgical action has the status of being at best unimportant, and at worst, through its ability to distract, actually harmful.

Ironically, this attitude came about partly through something Jones could only have applauded: the Second Vatican Council's positive emphasis on the way in which sacraments not only impart grace, but are also such that "the very act of celebrating them disposes the faithful most effectively to receive this grace in a fruitful manner, to worship God rightly, to practice charity."[28] This passage from the Constitution on the Sacred Liturgy reflects, as Mark Francis has rightly noted, "a new emphasis on the old scholastic dictum *sacramenta significando efficiunt gratiam* (sacraments cause grace by signifying)." Francis explains:

> While there is no dispute about sacraments being efficient causes of grace, the documents of Vatican II insist that attention also be paid to how the grace is communicated. . . . The reforms in Roman Catholic Sacramental Worship mandated by the Council were essentially attempts to help the celebration of the sacraments "signify" or communicate more effectively.[29]

Among the changes that this renewed emphasis on communication brought about was a new outlook on such things as bodily actions, ministerial clothing, the sprinkling of water, and the use of incense. These were either simplified, reduced in frequency of usage, or demoted to the status of optional "extras" because they were perceived, quite simply, as having hindered this communication. As Francis puts it, the attempt to help the sacraments signify more effectively was precisely the motive for stripping away "secondary celebrative elements that tended to overshadow the central action of many of the sacraments."[30]

This concern for the signifying capacity of the central aspects of liturgical action was undoubtedly valid. We must ask, however, whether the perception of the hindering effect of "secondary celebrative elements," to which this concern was allied, was accurate, or whether, on the contrary, stripping away so many of these elements has failed to heighten the signifying power of the central elements of liturgical action and has actually diluted that power. In the light of the sort of perspective that Jones helped to clarify (especially when this is supplemented by sociological insights), it is arguable that the affirmation of sign making that was communicated by "secondary elements" is precisely what made possible the effectiveness, humanly speaking, of the primary one.

By this I mean that an understanding of sign making as central to the nature of humanity is, as Jones recognized and bewailed, now rather rare. It is limited to what some sociologists would call a cognitive minority. From a sociological perspective, the maintenance of this minority requires that there exist, within that minority's common life, effective "plausibility maintaining mechanisms" that act at a "pre-theoretical" level. From this perspective, what we find plausible and meaningful is not so much the theoretical frameworks we hold to be true, but the social actions that symbolize and underpin these frameworks at this pretheoretical level.[31] In relation to the religious cognitive minority that is the church, this means that ritual and communally received narrative, for example, are as much a precondition of doctrine as an expression of it or a didactic aid. (The fact that specifically theological expressions of this sort of insight are to be found in "linguistic" understandings of religious language, such as that of George Lindbeck,[32] also may be significant here.)

Jones himself seems to have recognized an aspect of this when he complained about the tendency of the apologists of his own church to see the sacraments in didactic terms—as "helps to our 'infirm' condition rather than as absolutely central and inevitable and inescapable to us as creatures . . . whose nature it is to *do this*, or that, rather than *think* it."[33] But because the importance of the sociology of knowledge was only beginning to be widely recognized at the time of his death, he was unable to see that there is a sociological corollary of this kind of insight. That is, if a sacramental understanding depends on a pretheoretical acceptance of the efficacy of signs, then the plausibility-maintaining mechanisms of any sacramentally focused cognitive minority must reinforce belief in the general efficacy of signs with great vigor. Among other things, this arguably means that liturgical action must include a great deal of sign making of the sort that was common before the reforms but is now much less evident. From this perspective, the plethora and multilayered nature of traditional liturgical sign making are precisely what allows the central sign making of liturgical action to be effective.

This is not, of course, to say there should be no discrimination in the matter of "secondary elements." The potential dangers inherent in their use are well illustrated by Jones's own paintings, which often suffered, toward the end of his life, from what Jonathan Miles and Derek Shiel characterize as a clogging with information in which, "where everything is given uniform importance as a sign, the total significance collapses under the strain."[34] Nevertheless, just as Jones's best painting manifests a perfect balance between primary and secondary sign-bearing elements, so too the best liturgical practice seems to strike a similar balance.

All too often, however, by removing so much that is seen as secondary, current liturgical practice in the West has the effect of isolating the sacraments from the sign-bearing human context that makes them, anthropologically, an appropriate means of grace. If Jones was right in saying that "without *ars* there is no possibility of *sacramentum*," then just as in the secular world "works of art" have been isolated from the wider context of *poiesis* that he wanted to affirm, so also liturgical change has often insulated sacraments, in the technical sense, from the wider sign-making framework that makes them effective. It is not simply, as Jones put it, that "these blasted liturgists have a positive genius for knocking out *poiesis*."[35] In addition, once *poiesis* is knocked out in relation to secondary elements, it becomes ineffective in relation to the primary element that this knocking out was intended to reinforce.

In exploring this insight in the context of Jones's thinking, we must be very careful in our use of the term *sign*. As Miles and Shiel's otherwise excellent study of Jones indicates, it is all too easy to miss what is implicit in his use of the term. For example, when they note that Jones, as he grew older, "was agonized by what he considered to be a decline in sign-making," they immediately comment that this was in the context of "an explosion of sign-making on a scale that could not have been possible in a less technologically advanced society."[36] Perhaps, they suggest, Jones's "inability to respond to electricity instead of candlelight . . . resulted from his being born in a place and at a moment which he experienced as an interface between the country and the city, the nineteenth and the twentieth century."[37] The changing world order that Jones experienced meant, in their view, "a changing order of signs" that required "new evocations,"[38] and Jones's inability to recognize this was one of his chief limitations.

However, Jones's own perception that there was not "a changing order of signs" in this sense can surely be affirmed. The "explosion of sign-making" of which Miles and Shiel speak was, in fact, no more than an explosion of surface-level emblems of the sort that the advertising industry creates and manipulates. The fact that such emblems can occasionally transcend the context in which they were devised and acquire a deeper cultural significance need not be denied, of course. Nevertheless, the relative rarity of this transformation eloquently conveys the need to recognize the difference between signs, in the sense that Jones seems to have understood them, and contrived emblems.

To put it bluntly, some sorts of symbolism are effective because, for whatever reason, they seem "natural." (The Jungian, for example, will say that this is because of the way in which they relate to archetypes of the collective unconscious.) Other attempts at symbolism have no deeper hold on us, even if there is no obvious reason why this should be the case. Though we may not understand why, a lit candle *is* a more profound symbol than an electric lightbulb, just as the notion of Christ the shepherd of the sheep is more effective than that of Christ the managing director. The fact that we now live in an electrically lit, commercial society and not in a candlelit, pastoral one has simply not rendered the old imagery redundant or allowed an alternative new imagery to emerge.[39]

What this implies for liturgy is that, if we acknowledge the need for a Western restoration of sign-making activity over and above that which has been retained, this cannot be done effectively simply by introducing the sort of contrived symbolism that is sometimes invented by liturgical experimentalists. Though new symbolism may occasionally prove effective and should not be proscribed for experimental use, we surely must turn primarily to the tried, tested, and essentially *simple* sign making of the traditional liturgies of both East and West. As witnessed in the quality of the worship of the Taizé Community, the Fraternités Monastiques de Jérusalem, and Eastern rite Christians, the possibility of rediscovering liturgical *poiesis* lies primarily in things as rudimentary as verbal repetition, bodily action, contemplative chant, and the simplest of sensual elements: water, light, incense, oil, and the like.

Thus, even if David Jones may have been wrong about many of the specific aspects of liturgical reform that he lamented so vociferously, he might have been right in his more general and inarticulate sense that something was amiss. What is more, if he was at least partially right in thinking that the "blasted liturgists" of his own generation had "a positive genius for knocking out *poiesis*," important aspects of his thought suggest that a new generation of them might rather easily put it back in again. What is required, in this perspective, is not a major rewriting of the various Western service books, but simply an augmenting of their content by the sort of nonverbal sign making with which our forebears were familiar.

This will happen validly, however, only if it is motivated by a real understanding of the nature of signs. Jones's perspectives evidently go some way toward providing this understanding at an anthropological level. But they provide no way of seeing how the anthropological realities to which he points fit into the wider perspective of God's creative intentions. At this point, I believe, a pansacramental naturalism and certain aspects of Eastern patristic thinking must be brought into the picture.

The relevance of a pansacramental naturalism may be stated relatively simply, in terms of its belief that anything that has emerged naturalistically within the cosmos and tends toward an authentic apprehension of God should be seen theologically as an aspect of God's will. Thus, just as the aspect of human psychology that allows revelatory experience to occur must be seen as divinely willed, so also and more specifically, we must see as divinely willed the artistic element and archetypal orientation of that psychology. For this orientation is precisely what leads us to apprehend the cosmos in terms of that sense of "the architechtonic significance of the image in the created order," which Louth identifies in both Jones and John of Damascus, and which is implicitly at the heart of Christian thinking about Christ's incarnation. From this perspective, what some refer to as an "evolutionary epistemology," which recognizes that our understanding is limited by our faculties' evolutionary history, becomes an extremely liberating notion. Our limits, as much as our abilities, are in this perspective divinely willed.

The significance of Eastern patristic thinking here lies in its way of expanding on this insight. The most important aspect of this expansion may be the way in which the thinking of John of Damascus can allow us to expand Jones's notion of the importance of the human imagination. For John's notion of this importance is similar to that found in Jones, even though a stress of this kind is, as Louth has noted, "unusual in the Byzantine ascetic tradition, and in the Platonic tradition to which it is indebted."[40] While Jones cannot go very far beyond his conception of the use of signs being "natural to man . . . by virtue of his being an artist," John provides a way of beginning to understand, from a theological perspective, just what it is to think theologically "like an artist."

Seen from this perspective, the use of signs enables us to approach realities that will remain out of focus if approached only with the discursive reason. Responding to these signs involves not only what the Greek patristic tradition calls the logical faculty (*dianoia*), but also a higher faculty that is usually translated as the intellect (*nous*). This faculty is, however, far more than what we would now call the intellect, since through it, according to this tradition, God and the inner principles of created things may be known by a kind of direct apprehension or spiritual perception. It is the organ of contemplation, intimately linked to what the Makarian homilies call the eye of the heart.

This tradition is clear, however, that not all image production is derived from this level of direct apprehension. As some of the commentators on this issue have put it, one of the goals of the spiritual life is a kind of knowledge that "transcends both the ordinary level of consciousness and the unconscious." In this context, they say images "may well be projections on the plane of the imagination of celestial archetypes, and . . . can be used creatively, to form the images of sacred art and iconography." However, they go on, many of the images that erupt into human consciousness will not be of this kind. They will "have nothing spiritual or creative about them," since they "correspond to the world of fantasy and not to the world of the imagination in the proper sense."[41] Because of this, any imagery that arises as an aspect of individual experience may be spiritually dangerous. It needs to be subjected to *diakrisis* (spiritual discrimination), and for all but the most spiritually advanced, this is not something to undertake alone.

The former, positive aspect of the image-producing and sign-using faculties of the human psyche is the focus of John of Damascus's understanding. The latter, negative aspect is characteristic of much of the Byzantine ascetic tradition. This duality of focus is necessary, however, and underlines an important aspect of the psychological-referential model of revelatory experience that we have been examining. This model suggests, as we have seen, the possibility that the main faiths of the world have arisen through different kinds of contemplative experience, often manifested in visionary form. If understood simplistically, this may seem to imply that all visionary experience is genuinely revelatory of the divine reality. However, the negative, "fantasy" aspect of the human imaginative faculty is such that this cannot simply be assumed to be the case. The existence of this negative aspect means that at least some of the experiences at the heart of the faiths of the world may have no true reference to the transcendent reality that we call God, since they may be related to the world of fantasy and not to the world of the imagination in the proper sense. This possibility makes spiritual discrimination as important a factor in interfaith dialogue as it is in individual spiritual direction.

11

The Fallen World and Natural Law

As we have seen, a pansacramental understanding of the character of the natural world affects our thinking about God and God's relationship to creation in a number of ways. An aspect of this that we have not yet considered is that of ethics. This has many dimensions, not least, at the present time, that of its relation to environmental issues, which we shall consider further in chapter 14. In this chapter, however, I want to focus on something that has important ramifications in quite another area of ethical thinking and that will lead us into the discussion of the next chapter. This is the concept of "natural law," which is familiar to many because it is invoked by the Roman Catholic Church in its official teaching about artificial contraception, according to which it is wrong to indulge in the human sexual act in any way that deliberately frustrates its "natural" purpose: that of procreating children.

As a result of the practical difficulties of conforming to this ruling, many in practice ignore it, and at least some of these people attempt to justify this stance by claiming that the natural-law tradition is somehow fundamentally flawed. However, many of those who make this claim not only manifest little recognition of the subtlety and complexity of the discussions that have given rise to the official ruling, but also often ignore or belittle the essential insight on which the natural-law tradition is based: that Christian ethics cannot be based solely on "revealed" truths, but necessarily requires at least some degree of philosophical underpinning.

For the natural-law tradition, this underpinning is provided at least partly by reflection on God's intentions as the world's creator, and in this aspect of its approach, it is, in my view, valid. In practice, we often face complex ethical decisions that cannot be answered in terms of a simple reading of God's self-revelation in historical acts. Therefore, if we believe—as I do—that theologians working within this tradition have failed to complete the task of making explicit God's creative intentions as they apply to ethical problems, this is simply a challenge to us to address anew the task that they have undertaken. In this chapter, I shall suggest that this not only is possible, but also can be done at least partly in terms of the perspectives on the created order that we have already examined.

My starting point is the observation that defenders of the natural-law tradition frequently manifest a rather narrow view of how God's creative intentions should be read off from the empirical world. The problem is not merely that the scholastic reliance on classical legal thinking may have led to an undue stress on human faculties at the expense of the human person or community, though this is certainly arguable.[1] More important, much of the recent practice of natural-law thinking has been based (as I shall argue in what follows) on a view of "nature" that requires expansion from theological and scientific perspectives.

Theologically, this argument may be approached through an important aspect of the Eastern Christian tradition: seeing the empirical world not as natural but, in a theological sense, as subnatural. Philip Sherrard, for example, has expressed this view in terms of the different understandings of the resurrection of the dead that were expressed in the patristic period. In the West, he notes, there was in this period a strong stress—as in Tertullian—on the resurrection of the "flesh," by which was meant quite explicitly the flesh of the body experienced in our earthly life. In the East, by contrast, there was an alternative understanding, especially in writers such as Gregory of Nyssa and Maximos the Confessor. In this understanding, says Sherrard, the resurrection body was not identified with the body in its present state, "composed of juices and glands and organs for excreting and procreating, and subject to the processes of conception, childbirth, adulthood, old age, sickness, and death." These aspects of the earthly body were not seen as parts of the "original" body intended by God when God created the world. They were, rather, "aspects of the 'garments of skin' added to the original body as a result of the fall. They are as it were accretions, things grown over the body."[2]

This early divergence between East and West—symbolized by the Eastern allegorical interpretation of the "garments of skin" given to Adam and Eve (Gen 3:21), to which Sherrard alludes—may be seen as emblematic of a more general and continuing divergence in relation to the theological understanding of the empirical world. On the one hand, by its stress on the effects of the Fall, the Eastern Christian tradition has tended in many respects to be more pessimistic about the empirical world than has the Western tradition. Paradoxically, however, it has in other respects been more optimistic. Despite its stress on the "fallen" character of the empirical world, it has, as we have seen, tended to avoid the notion of some kind of "pure nature" to which grace must be added as a supernatural gift. Rather, as Vladimir Lossky has put it, there is, for the Eastern tradition, "no natural or normal state, since grace is implied in the act of creation itself."[3]

For Sherrard, this more optimistic strand is reflected in the view of the human condition that the Eastern tradition has tended to take. Western theology, he says, conceives of humanity—even as originally created in God's image—as "a union of the animal or organic life and the intellectual life. The animal or organic life is not [as in the East] seen as superadded to man as a consequence of the fall." In addition, he goes on, the spiritual aspect of human nature is not, in the West, seen as an intrinsic part of the human condition. "On the contrary," he says, for Western theology it is "the spiritual life which is added to man's natural state. Man is not spiritual by nature, as he is in the Eastern Christian tradition. He is spiritual through a supererogatory act of grace."[4]

When Sherrard speaks in these terms, it is perhaps arguable that he oversimplifies the cosmological and anthropological understandings of East and West. Certainly, he seems to have little sense that the tendencies he contrasts may represent complementary rather than antithetical ways of understanding the created order. Despite this weakness, the contrast that he emphasizes arguably remains useful for the purpose for which he makes it: that of exploring the theologies of marriage and sexuality that are either implicit or explicit in the two parts of the Christian world.

At one level, says Sherrard, the issue here may be expressed in terms of the question of what it means to speak of marriage as sacramental. In medieval Western thinking, he claims, not only was this notion limited very largely to the symbolic link between marriage and the union of Christ with the church, but also the understanding of this symbolism "was limited in a manner which prevented a full realization of its scope."[5] A sacred symbolism, he goes on to explain, "becomes a creative or spiritualizing influence when it is seen as capable of acting upon the matter to which it applies in such a way that it helps to transform this matter into the reality which the symbolism is intended to signify. This presupposes the perception that within the matter to which the symbolism applies there is the capacity or the potentiality to be transformed in this way."[6]

For Sherrard, this means that only if it is "understood that the relationship between man and woman is capable of possessing an eternal and metaphysical character can it actually become a fully achieved sacramental union." Neither in the thought of Augustine nor in that of the later scholastics, however, "is there any recognition that the relationship between man and woman is capable of attaining a sacramental dignity in this sense." Indeed, he goes on, nowhere in the medieval West was there any doctrine

> in which sexual love is recognized as providing the basis of a spiritualizing process whose consummation is the union, soul and body, of man and woman in God, a revelation of the divine in and through their deepening sense of each other's being. The idea that the sexual relationship might create a metaphysical bond which death itself is powerless to destroy is alien to the mind of medieval theology as a whole. Marriage is regarded above all as an ecclesiastical or social institution designed for procreation.[7]

Against this historical background, recent thinking within the Roman Catholic Church represents, in Sherrard's view, an advance. This, he says, is reflected even in the encyclical letter *Humanae Vitae*, in which the teaching against artificial contraception is reiterated. The document, he observes, clearly sees marriage in terms of its status as the gift of husband and wife to one another, a union through which they may perfect one another. However, he goes on to say:

> This positive and enriching image of marriage is not enlarged on or even allowed to stand in its own right. It is made subordinate to the conventional non-sacramental view of the early theologians: that the principal end of marriage and that which uniquely specifies its nature is the procreation and education of children. We are told in effect that the perfection of each other which man and woman may achieve through marriage is

not an end in itself, but exists "in order to co-operate with God in the procreation and education of children." . . . This is the ultimate purpose of marriage, its final *raison d'être*. It is not that through their union man and woman should achieve the integrity of the human creature by means of an inner transformation of the mortal and corrupt conditions of their present existence.[8]

A major aspect of the problem here is, for Sherrard, that when nature is spoken of in the document, it is effectively identified both with what God has willed and with "nature in its present state, not as it is in its original state, as it issued from the hand of God." Here, he goes on, we are "within an order of theology which represents an uneasy alliance between the [Augustinian] conception of original sin . . . and Aristotelian optimism in respect of mundane existence." The effect of this alliance, in his view, "is that to all intents and purposes the event described as the 'fall of man' is treated as something that conforms to the will of God, and consequently there is no sharp distinction made between the order of nature prior to the fall and the order of nature subsequent to the fall; both are treated as expressing the will of God."[9]

In this way, Sherrard goes on, Western theology tends to see humanity's fallen life and the natural processes to which humans are subjected in the fallen world as expressing the will, pleasure, and purpose of God and thus "as constituting the norm on which the moral law of the Church is to be based." However, he argues, if it is understood "that it is not the fallen state of nature and of man which is natural, but their pre-fallen and paradisaical state, and that it is this state which expresses the will of God, then a quite different attitude to the relationship between the moral and natural law will prevail, and quite different conclusions may be drawn from it as a consequence."[10]

At this point, perhaps, we can begin to perceive the way in which the views of Sherrard may be somewhat one-sided, in a way that goes beyond mere polemic or historical oversimplification. When he says the world as we know it "is not that which God has created or intended for man, but is what man has brought on himself as a result of his own defection and error,"[11] there seems to be little recognition that the consequences of the Fall should be seen not simply as a retributive punishment, but also as a divinely ordained "medicine." In fact, here and in other parts of his study, something in Sherrard's tone—if not in the formal content of his argument—suggests that his attitude to the empirical world, while far from puritanical in its effect, has parallels with the sort of dualism that is inherent in a puritan position.

As we shall see in the next chapter, however, this tendency is not intrinsic to the Eastern tradition on which Sherrard attempts to base himself. Within that tradition, there may certainly be writers who manifest a similar tendency. (He himself cites the views of Gregory of Nyssa and Maximos the Confessor on human sexuality as reflecting a flawed anthropology of a sort that may be interpreted in this way.[12]) In these writers, however, a characteristically Eastern stress on the presence of God's *Logos* in the world provides, as we shall see, a sort of counterbalance to their otherwise perhaps dualistic instincts.

Can we then say that Sherrard, because of his tendency toward a dualistic view, is simply mistaken in his critique of the Western natural-law tradition? Certainly, I believe, we can see his critique as incomplete and in certain respects distorted. Despite this, however, we can surely see his main point as at least arguable: that we should not attempt to understand God's creative purposes only in terms of the empirical world. Our understanding must also take fully into account our eschatological hope, our understanding of the state to which we are, in our journey through this world, being led. Sherrard's perception that this world represents a journey of this kind—one that may be properly understood only in terms of God's ultimate intentions for us—is surely one from which we have much to learn, and not least in relation to his reflections on marriage and on human sexuality.

In particular, especially in the light of contemporary psychological insights, we can learn a great deal from Sherrard's strong sense of the sacramental potentiality and divine origin of the deepest roots of human sexuality. If, as we surely should, we accept his view that the sexual energy in man and woman "has its source in the deepest strata of their life . . . is rooted in the ultimate mystery of their being . . . [and] is the source and generator of all human creativity," then much will follow. In particular, it will be no great step for us—especially if our faith is incarnationally focused—to acknowledge that this energy, as he goes on to say, "derives its sacramental quality not from any purpose, such as the procreation of offspring . . . [but] from the fact that its own origin is divine and its own nature is sacred."[13]

Before we can adopt this position, however, we need to understand one aspect of Sherrard's approach more fully, since it will strike a note of dissonance if it is understood in a superficial way. This is his way of pointing toward God's ultimate intentions largely in terms of the "pre-fallen" state of the cosmos. Given the scientific evidence that the empirical world has never undergone a transition of the sort he seems to envisage, it may at first seem that a stress on some hypothetical prefallen state is now no more than an anachronism.

We must remember, however, that Sherrard—as his reflections on the resurrection body indicate—is not merely looking backward when he speaks of our prefallen state. His focus on that state is always implicitly used to point us toward our eschatological hope. Moreover, as we shall note further in the next chapter, using scientific evidence of the world's continuity as a counterargument to Sherrard's approach ignores the subtlety of his concept of the Fall. For he does not see the Fall as occurring within the empirical space-time processes that the scientific methodology can investigate. He is speaking, rather, about what he calls a fall "*into* a materialized space-time universe."[14] In this context, talk of a prefallen state is misleading if the *pre-* suffix is interpreted in terms of empirical temporal duration. Because of this, the sciences—which deal only with this empirical dimension of reality—can have nothing to say about the matter. By definition, they deal only with the fallen world.

This is not, however, to say that the sciences have nothing to teach us when we reflect on the natural-law tradition (though Sherrard himself, with his strong antiscientific bias,[15] would probably have taken this view). One of the main problems of medieval thinking about natural law was that, given the science of the time, it inevitably assumed an essentially static created order. The purposes of God in this order were thus believed to be simply readable (so to speak) from the way things were seen to be. In the evolutionary perspective of modern scientific theory, however, this simple kind of "givenness"—still assumed by the vast majority who defend the concept of natural law—can no longer be taken for granted. To understand why things are as they are, we now recognize that we have to understand both how they have been in the past and why they have developed in the way that they have.

Thus, for example, for the old static model, sexuality could be understood in terms that partly relied—if only implicitly—on the analogy with other animals. The physical structure of the human sexual act was clearly identical to that of other mammals, and this encouraged the belief that its purpose was therefore identical in the two cases. God's intention for human sexuality could thus easily be seen in purely procreative terms. In an evolutionary perspective, however, this approach is no longer tenable, at least in any simplistic way. The concept of evolutionary adaptation applies not only to physical characteristics, but also to behavioral ones. Physical organs and behavioral traits, we now realize, can and do take on new dimensions of significance and purpose during the evolutionary process.

For the Christian who accepts these basic scientific insights, this is immensely important. If we see God's creative will as being worked out through the developmental processes to which the scientists point, then we will see significance not only in the created order as it is today, but also in the *direction* of the evolutionary processes that have given rise to the situation in which we find ourselves. In particular, the later developments of the evolutionary process will often be the ones that we tend to see as defining God's prime intention for us now, especially when these represent what differentiates us from the rest of the animal kingdom.

For our remote, prehuman ancestors, for example, the sexual act was, as for other animals, simply the result of an instinctive urge that led to the divinely willed result of reproduction. But by the time God had produced human beings through the evolutionary process, new and far more complex aspects of the sexual dimension of our lives had arisen. It is these on which many now tend to focus when they speak about God's intentions for our sexuality, and arguably this focus is theologically justifiable.

For example, one aspect of this human evolutionary development has been an increasing stress on the sexual dimension of life. (This stress is now, in fact, far greater than that usually found in other animal species,[16] for most of which there has apparently been no evolutionary pressure to indulge in sexual activity more than is required for the reproduction of the species.) One reason for this increased emphasis among humans has presumably been the fact that our evolutionary development has involved the unfolding of something that is largely absent from the reproductive behavior of other animals: the kind of tenderness that now usually accompanies the human sexual act, whose forms, "such as kissing and caressing, are in origin actually actions of parental care."[17] Yet another (and again probably related) development has been—as Sherrard has stressed in the wake of numerous psychologists—the emergence of a strong link between sexuality on the one hand and creativity and religious responsiveness on the other.[18]

All of these, if we take seriously the notion that God has worked through the evolutionary process, may be seen as signs of the way in which God can and does take our rootedness in our animal nature and mold it for God's ultimate purposes in us. The sexual developments that have occurred during the later stages of our evolution can be understood in terms of the "emergent properties" that many—especially within the current science and religion debate—stress as a way of speaking about God's action through natural processes.

Theologically, this aspect of the evolutionary process may, interestingly, be linked to the kind of perspective we have noted in Sherrard's arguments. In particular, as we have already seen (and will see in a new way presently), this may be done through the strand of Greek patristic thinking that, when speaking about the way in which the *Logos* of God "became flesh" (John 1:14), stresses the way in which everything was, in the beginning, created through this *Logos* (John 1:1–4).

Even if this particular approach is not accepted in detail, however, my more general arguments about the evolutionary perspective will remain valid, as will the other points I have made. Thus, the question with which we began—that of the ethical implications of modern contraceptive practice—may be seen as not yet having been answered in a definitive way. Recognition that aspects of the Eastern Christian tradition and of modern scientific insight provide a whole new dimension to the debate does not, admittedly, lead to the assumption that all previous conclusions must be reversed. It does, however, suggest that the conclusions that have so far been drawn can be seen, at best, only as provisional.

12

The Fallen World and Divine Action

The Eastern patristic interpretation of the biblical story of the "garments of skin" given to Adam and Eve after the Fall (Gen 3:21) has, as we have seen, vital implications for reappropriating and expanding the natural law tradition. It also, as we shall see in this chapter, has wider ramifications, especially for the pansacramental naturalism I have advocated. For this reason, it is important to note that when Eastern writers spoke of these garments of skin, they were not simply developing a theological model on the basis of the kind of allegorical biblical interpretation that many Christians now find questionable. Rather, we may see this term as a kind of shorthand for a more general and easily accepted insight: that the biblical distinctions between the pre- and post-Fall condition of humankind, and between this world and the "new creation," are such that we must be careful in using our experience of this world as a guide to God's original or ultimate intentions for the created order. Western theology tends to err in its reflection on creation, according to this viewpoint, precisely because it fails to recognize the reality to which the notion of "garments of skin" points, and this remains true even if this phrase and its traditional allegorical interpretation are set aside.

An important aspect of this Eastern approach is linked to the philosophical discussion of what, in the West, is usually termed the problem of evil, which arises when our conception of God—in which God is understood to be omnipotent and wholly good—is coupled with the observation that evil exists. As many have noted, this combination seems to involve an inherent contradiction.[1] One aspect of this apparent contradiction is, admittedly, relatively easy to solve: when faced with the question of why God allows "human" evil, we tend to be convinced by the argument that human free will is so central to the divine purpose that hindering it can only rarely, if ever, be appropriate. When it comes to the problem of "natural" evil, however, most of us are less sure in our response. Faced with some of the unpalatable consequences of the laws of nature—the child drowned by a tidal wave, for example—we cannot help feeling uneasy. We may recognize, with F. R. Tennant and others, that rational choice is possible only against a background of regularity of a kind that the laws of nature provide.[2] Nevertheless, we cannot help wondering why a created order that God saw as "good" is not more uniformly benign than it is.

For some, an answer to this problem lies in the occasional biblical hints about the universe being, in some sense, under the sway of "principalities and powers" that are hostile or indifferent to God's ultimate intentions (e.g., Eph 6:12; Col 2:15). For many Western Christians, however, this concept has proved difficult, both because of its mythological imagery and because it has seemed to contradict the notion that God made the world and saw that it was "good" (Gen 1:25, 31). It is arguable, nevertheless, that this principalities-and-powers notion is considerably important.[3] Even if we regard it largely as a mythological concept, with no straightforward referential content, it does clearly point toward an important aspect of the tension inherent in the Judeo-Christian notion of what it means to speak about the empirical world as God's creation.

Put in biblical imagery, this tension might be expressed as follows: On the one hand, we do, with the psalmist, want to say that the lion stalking its prey is seeking its food "from God" (Ps 104:21). On the other hand, we also want to say, with the book of Isaiah, that when God's ultimate purposes are fulfilled, "the wolf shall live with the sheep, and the leopard lie down with the kid" (Isa 11:6). And in wanting to use both these images, we are implicitly admitting to an ambiguity in our attitude to God's creation as we experience it. We want to say both that this is God's world and that it is not yet *fully* that world as God ultimately intends it to be. And as the animal imagery indicates, the problem of evil—at least as the carnivorous hunter's prey might perceive it—is an important aspect of this tension.

At this point, it seems to me, Western philosophical approaches to the problem of evil ultimately fail to go far enough. They may be correct, for example, when they suggest that human life and choice are possible only against a background of regularity that necessarily has unwanted effects. They may also be correct, as John Hick's creative development of Irenaeus's thought suggests, that evil is a necessary aspect of a world that is a "vale of soul-making."[4] In their different ways, however, such approaches assume that our understanding of the link between this world and the "new creation" can be developed only in terms of the question of how the problematic aspects of the former may be seen as necessary to provide conditions through which the latter may be attained. It is assumed that God's will can in some sense encompass the evil in the world because a greater good is derived from it.

While this is little more than an assertion for the Western theological approach, the Eastern Christian tradition has resources here that Western traditions have, at best, only in a diluted form. These may be expressed in terms of the way in which the effects of sin are seen as inhibiting the purposes of God within the created order but not as destroying God's image in it. In a fallen world, there is still, for the Eastern Christian mentality, a teleological movement in which all things—even sin itself—are taken up and used for the divine purpose.

This has been expressed in a number of ways,[5] most interestingly, perhaps, by a notion of Maximos the Confessor about the way in which the underlying principles of all created things have a vital role in leading those things to their intended final end. But before we examine this aspect of Maximos's thinking, it is important to explore the more general Eastern framework within which his ideas are set: that which has at its center the notion of the "garments of skin," which we have already noted in the thinking of Philip Sherrard and which, as we have seen, recognizes that the world we experience does not correspond to God's ultimate intentions.

The best modern summary of this patristic notion is perhaps that of Panayiotis Nellas, who begins by stressing the way in which humankind is seen biblically as having been made in the image of God (Gen 1:26). This notion, says Nellas, provides for the fathers of the Eastern Church the basis for a perspective in which "the essence of man is not found in the matter from which he was created but in the archetype on the basis of which he was formed and towards which he tends."[6]

This belief—that the ontology of humankind is not found in matter, but in the image in which humans were created—does not, Nellas stresses, mean we should interpret this patristic picture in simple dualistic terms, in which the "soul" is what was created in the image of God. Rather, he says, this belief is based on the following perception:

> [The] ontological truth of man does not lie in himself conceived as an autonomous being—in his natural characteristics, as materialist theories maintain; in the soul or in the intellect . . . as many ancient philosophers believed; or exclusively in the person of man, as contemporary philosophical systems centred on the person accept. No: it lies in the Archetype. Since man is an image, his real *being* is not defined by the created element with which the image is constructed. . . . Man is understood ontologically by the Fathers only as a theological being. His ontology is iconic.[7]

What, then, is this "Archetype" that Nellas sees as so central to Eastern patristic thought? Fundamentally, he says, the key lies in the prologue of the Fourth Gospel (John 1:1–14): "the archetype is Christ,"[8] not simply as the *Logos* of God, but more specifically as the *incarnate Logos*.[9] This stress on incarnation, we should note, reflects precisely the strand of thinking that we have noted in the views of Philip Sherrard, since the notion of incarnation is, for Nellas, as applicable before the historical incarnation in Jesus as after. On the level of the "supra-temporal reality of God," he insists, "Christ is 'the firstborn of all creation' (Col 1:15–17)"[10] and therefore "is not a mere event or happening in history." Even before the Fall, says Nellas (reflecting a view found in Irenaeus and others), man had "need of salvation, since he was an imperfect and incomplete 'child.' "[11] Christ accomplishes man's salvation "not only in a negative way, liberating him from the consequences of original sin, but also in a positive way, completing his iconic, prelapsarian 'being.' "[12]

In this perspective, Nellas continues, the history of humankind—and indeed of the whole universe—can only partially be understood in terms of secular disciplines. "Since the ontological origin of man is not to be found in his biological being," he says, "but in his being in Christ, and the realization of his being in Christ constitutes a journey from the . . . iconic to that which truly exists, history can be understood precisely as the realization of this journey. As such, it has its beginning and its end in Christ." Reflecting the notion of the cosmic Christ to be found in the biblical letters to the Ephesians and to the Colossians, the Fathers insist that "it is not only the present and the past which move and determine history but also . . . the advent, at the end of the age, of Christ the recapitulator of all things, that is, of the Logos together with His body, the transformed world."[13] This perspective, Nellas continues, implies:

> The development or evolution of humanity is illuminated inwardly. Our understanding of humanity is not determined simply by the processes of change which are observed in the matter of the image, but, without this first aspect being overlooked, our viewpoint is extended and understood primarily in terms of an evolution or raising up of the image to the Archetype. . . . Evolution in this way is understood in all its dimensions—not only in those which are determined by scientific observation—and is given its true and full value.[14]

Having set out the background in this way, Nellas next tackles the question of the nature of the "garments of skin." The notion of the empirical world as "unnatural" is, he says, central to "the teaching of the Fathers on human nature," which "forms, as it were, a bridge with two piers, the first pier being the understanding of what is 'in the image,' the second the deeply significant notion of 'garments of skin.' " These garments of skin are to be interpreted partly in terms of what is necessary for survival in humanity's postlapsarian state, but also partly in terms of the need to foster in a more positive way humans' "return to what is 'in the image.' "[15]

Intrinsic to the notion of the garments of skin, says Nellas, is the notion of mortality. The Fall, though in one sense a fall into materiality, is not to be identified simply with a fall into created matter. According to Gregory of Nyssa, for example, although the body has become "coarse and solid" through the Fall and is characterized by a "gross and heavy composition," it will at the resurrection recover its prelapsarian state, being "respun" into "something lighter and more aerial." The body will not be left behind, as a Gnostic dualism might maintain, but will be transfigured into its original beauty. Moreover, Nellas notes, Gregory does not consider that only the body is in need of this transformation. The functions of the soul also must undergo a transformation, having become "corporeal" through the Fall.[16] Thus, while Gregory, more than most patristic writers, may seem to identify the garments of skin with the postlapsarian human body, he is, according to Nellas, actually "referring to the entire postlapsarian psychosomatic clothing of the human person."[17]

Nellas sees the implications of this view most fully worked out in the work of Maximos the Confessor. Although Maximos's work is "so dense and so rich in different layers of meaning" that Nellas feels his own interpretation of it to be somewhat provisional,[18] he is confident in seeing as central to it the biblical statement that it was God who gave the garments of skin to humanity (Gen 3:21). This, says Nellas, is interpreted by Maximos in terms of the way in which "God acts in a loving way even to those who have become evil, so as to bring about our correction."[19]

The point here, says Nellas, is that although at one level the garments of skin are an evil, brought about as a direct result of human rebellion against the divine intention, God "changes that which is the result of denial and is therefore negative into something relatively positive." Therefore, he goes on, the garments of skin are "a second blessing to a self-exiled humanity." God has added this blessing "like a second nature to the existing human nature, so that by using it correctly humanity can survive and realize its original goal in Christ."[20] Thus, for example, God, by "allowing man to dress himself in biological life, the fruit of sin . . . redirected death, which was also the fruit of sin, against biological life, and thus by death is put to death not man but the corruption which clothes him."[21]

Moreover, says Nellas, such obvious evils as death are not the only aspects of our present life that can be seen in terms of this understanding. Following John Chrysostom, he relates the garments of skin to human work, the arts and sciences, and politics. In these aspects of human life, he says, we can see particularly clearly how the garments of skin "are not unrelated to the iconic faculties of man before the fall." God, he goes on, has enabled "the attributes of that which is 'in the image'—the attributes which were transformed into 'garments of skin' without being changed in essence—to be useful to man not only in his struggle for mere survival but also as a means of making the new journey towards God."[22]

Another aspect of the garments of skin that Nellas discusses is their relationship to the more general cosmic ramifications of the Fall. Here, it must be said, he is perhaps a slightly less reliable guide than he has been hitherto, since his emphasis on later Byzantine thinking leads him to incorporate perspectives in which the earlier patristic thinking of the East has been modified by Western concepts.[23] This factor does not, however, detract from his main point, which is that the Eastern patristic tradition does not see the Fall as having effects on humanity alone, but on the whole created order. The laws that govern that order, while they continue to operate after the Fall, are seen as doing so in a way that allows what the West calls "natural" evil: that is, they operate "in a disorganized and disordered way, and they involve man too in this disordered operation with the result that they draw him into misery and anguish."[24]

However, Nellas explains, according to the Eastern patristic tradition, these cosmic implications of the Fall have also been transformed by God. Like the garments of skin themselves, they constitute not only a penalty but also a remedy.[25] Because of this, even though some writers within this tradition come perilously close to a dualistic distaste for the empirical world, they express themselves in a way that manages, ultimately, both to preserve the Hebraic notion of the goodness of the world and to point to the way in which the problems inherent in some of its aspects may be resolved. The empirical world is, for this approach, appropriate to our fallen state, not only because its relative opaqueness to God's will reflects that state, but also because it remains sufficiently transparent to the divine intention to lead us toward the "new creation."

We must ask whether this vision can still be relevant to the concerns of Christians of the present day. At one level, it clearly has its attractions; as we have noted, it can subtly and powerfully supplement the kind of perspective on the problem of natural evil that has been developed by Western Christians in a more purely philosophical manner. At another level, however, a problem clearly arises: Can we really give any significant weight to an understanding that is tied, apparently inextricably, to a notion of the Fall as a historical event? Does our scientific understanding of the world—which seems to argue so clearly against the historicity of such an event—simply mean that this whole understanding must now be abandoned? How, for example, can we understand within this framework the existence of "natural" evil in the form of diseases evident in the fossil remains of dinosaurs, which lived millions of years before the evolutionary emergence of humanity?

Faced by this kind of question, it is important to recognize that, in the Eastern patristic tradition, the Fall is often seen as being, in some sense, what Philip Sherrard calls a lapse "*into* a materialized space-time universe."[26] There is a strong sense, especially in those influenced by the Origenist tradition, that the Fall occurs from a state that is outside our empirical temporal duration and our present biological state.[27] Therefore, it is not a "historical" event at all in the ordinary sense of the term. Nellas himself, as we have seen, speaks in another context of "the supra-temporal reality of God,"[28] and if we talk about a "pre"-lapsarian state of humanity, then it may well be necessary to speak about it in comparable supra-temporal terms. In this sense, the sciences can have nothing to say about the matter, since they deal only with the space-time universe.

This does not mean, it should be noted, that we have to abandon altogether the notion of a fall within the ordinary temporal process, since clearly the supra-temporal fall envisaged in Origenist thought may be seen as reflected and manifested in the emergence of a knowledge of the ambiguity of self-centered behavior in our early ancestors. From a purely scientific perspective, this emergence was clearly a feature of some stage in our evolution, and though we may not be able to say precisely when the first decision to "rebel against God" was made within our present space-time universe, we cannot avoid accepting that such a decision must, at some point, actually have been made. Thus, even if we feel the need to speak of a gradual emergence of a moral sense, and of "Adam's" decision to rebel against God as being constituted in history by many individual decisions, we can still affirm the historical basis of the Genesis story, provided that we see the cosmic "consequences" of this rebellion as occurring before its historical manifestation. This is possible if we can see this historical manifestation as having been, in some sense, "anticipated" in God's ordering of the creation, and this is certainly not incompatible with a supra-temporal understanding of an Origenist kind.

Whether or not we find either or both of these ways of thinking helpful is, however, irrelevant to my main point, which is that, as Nellas himself indicates, the notion of the Fall points accurately to central aspects of our existential experience. The situation of frustration, meaninglessness, and evil in which we find ourselves truly exists, and from this situation, God has, according to Christian belief, undertaken to rescue us in and through Christ. In this sense, our present situation—however we think it came about—is not a "natural" situation at all, if we use that term, as the Eastern patristic tradition does, to refer to God's "original" intentions. The notions of our "unnatural" state and of the garments of skin are therefore relevant to us, even if we have major reservations about the particular way in which they were originally formulated. In particular, in the context of our current exploration, they seem to have much to teach us about the relationship of "this world" to "the world to come."

Thus, for example, the Christian ascetic tradition becomes comprehensible when looked at in this perspective. When this tradition speaks of battling with the "flesh," this is not to be understood as battling with the material creation, since even after the Fall, this is, as we have seen, in many respects a "second blessing." The term *flesh* connotes, rather, the fallenness of our state. For instance, the "works of the flesh," as listed by Paul in his letter to the Galatians (5:19–21), include such things as envy and selfish ambition, which have no special connection with the body as we understand it. He thus uses the term *flesh*, as did many later Eastern patristic writers, to denote our "unnatural" psychosomatic structure, with which we battle, not in order to escape the body, but in order to attain its "natural" state.

The Eastern ascetic tradition is therefore not ultimately one of dualistic distaste for created things (though it is sometimes expressed in terms that can be interpreted in this way). It is, rather, one of "training" oneself to use these things properly, so that we can, as Olivier Clément has put it, "pass from a state 'contrary to nature' to a state 'in harmony with nature', in harmony, that is, with that human (and cosmic) material united in Christ with the godhead, without separation or confusion."[29] This understanding of ascetic struggle is, in fact, part of a considerably sophisticated psychological awareness in the Eastern patristic tradition, aspects of which bear marked similarities to the insights of the various modern schools of depth psychology.[30]

Still, what is most important about this perspective on the character of the empirical world does not lie in considerations such as these, or even in the light it throws on the problem of natural evil. What is most important, at least in the context of our present study, is its way of illuminating our understanding of how God acts in the world. For the Eastern Christian tradition, while sometimes speaking about God acting in a way that is "above" nature, has tended, because of its understanding of what constitutes the natural world, to avoid the technical distinction between the natural and the supernatural used in the West. Rather, because of Eastern theologians' view that there is "no natural or normal state [as envisaged in the West] to which grace must be added,"[31] they have tended to think about divine action in a far more subtle (if usually less systematic) way than Western writers.

This may be illustrated by the question of how God acts in and through the sacraments of the church, since for Eastern theologians, a sacrament is not only, as for their Western counterparts, an outward and visible sign of an inward and spiritual grace. It is also something more: what Alexander Schmemann calls "a revelation of the genuine *nature* of creation."[32] Philip Sherrard, in particular, has stressed this aspect of the Eastern patristic understanding, noting that a sacrament is not "something set over against, or existing outside, the rest of life . . . something extrinsic, and fixed in its extrinsicality, as if by some sort of magical operation or *Deus ex machina* the sacramental object is suddenly turned into something other than itself." On the contrary, he goes on, "what is indicated or revealed in the sacrament is something universal, the intrinsic sanctity and spirituality of all things, what one might call their real nature."[33]

What makes the sacrament necessary, Sherrard goes on to explain, is simply the way in which the Fall has led to the created order's "estrangement and alienation from its intrinsic nature." In the sacrament, he says, "this divided, estranged and alienated state is transcended," and the created order's "essential and intrinsic nature is revealed."[34] This means, he continues, that the sacrament is "reality itself, as it is in its naked essence and without anything being changed or symbolized or substituted." Therefore, terms as such as *transubstantiation* or *transformation* "tend to lead to confusion," since at the deepest level "nothing need be transubstantiated or transformed." The sacrament is a transformation only insofar as it is "a re-creation of the world 'as it was in the beginning.' "[35]

In this understanding, a sacrament is simply a manifestation of the true reality of some aspect of this world, so that its usual relative opaqueness to God's purposes for it and presence within it gives way to a complete transparency. The possibility of this manifestation is, moreover, not tied to a particular set of ecclesiastical sacraments. Although the Eastern Church has, like the Roman Catholic Church, often spoken of seven sacraments, of which baptism and the Eucharist are the chief, it has never in fact set out a definitive list of them. This is understandable, according to Sherrard, because "everything is capable of serving as the object of the sacrament," and their number cannot be fixed. If we must speak of "particular" or "greater" mysteries, this simply means we recognize "a sacred hierarchy of mysteries established in view of the particular conditions of individual existence in the world."[36]

This notion of the potential for any part of the created order to become more fully transparent to the purposes and presence of God is extremely important when we come to consider God's action in more general terms. While the created order evidently has a certain transparency to the purposes of God before any specific human invocation of divine grace, it is clearly not fully transparent to those purposes before that invocation. Certainly, the way in which the universe has evolved naturalistically does already indicate some degree of transparency to the divine purpose, as does the fact that its beauty and awesomeness can lead us directly to praise God as its creator. But as the problem of natural evil indicates, this transparency is only relative. In a "fallen" world, there is a certain opaqueness to God's purposes, and—as the centrality of intercessory prayer to the Christian tradition indicates—this is usually overcome, and God's "special" providence brought about, only through the human recognition and invocation of God's will.

From this perspective, divine action may be illuminated by the nature of the sacraments in a profound way. Quite simply, what is usually called special providence may be seen not—as Western theologians have tended to think—as the product of some kind of divine interference with the world, but rather as the outward manifestation in this world of something that is already present but hidden within it: what we can properly call its "natural" state. The miraculous is not, in this perspective, the result of something being added to the world. It is, rather, the wiping away from that world of the grime of its fallen state, in order to reveal it in its pristine splendor.

Thus, in this perspective, although the old distinction between general and special providence in terms of different modes of God's action becomes superfluous, the terms remain meaningful at another level. Their meaning lies not in the traditional Western distinction between modes of divine action, but in a distinction between different degrees of human response to the divine will. What is called general providence corresponds to the aspects of the world that are, independently of the human response to God, still "transparent" enough for God's purposes to be fulfilled. These will include, as we have seen, all the naturalistic and scientifically explorable processes that have allowed the evolutionary emergence of beings who can respond to God in faith. What is called special providence corresponds to the aspects of the world that are, so to speak, inoperative until this response in faith is made.

In this perspective, when the universe "changes" so as to bring about events of special providence, it is a sign and a foretaste of what is to be when all the purposes of God have been fulfilled. In such an event, created things are, in the deepest sense, simply becoming themselves as they are in the intention of God. As the grime of fallen human nature gets wiped away in any person through response to God in faith, not only is the fullness of human potential, which we see in Jesus, actualized in that person to some degree, but also the world around that person may be cleansed, so that nature becomes "natural" once more. In this perspective, special divine providence and human sanctity are inextricably linked, so that it is no accident that anticipatory experiences of "the wolf lying down with the lamb" are linked in the memory of the Christian community with the responses of wild animals to people like Francis of Assisi, Cuthbert of Lindisfarne, and Seraphim of Sarov.

In this way, I believe, we can see from a theological perspective the appropriateness of speaking of miracles. They are not, for this perspective, the result of divine interference with the world. Rather, they are reflections of an aspect of the *true* nature of the world, which is usually hidden from us, and in this sense, they may be seen as manifestations of "laws of nature" that reflect, more fully than those laws of nature that are scientifically explorable, God's presence in all things. In fact, this factor is what links this whole perspective to the kind of pansacramental naturalism I have defended in the earlier chapters of this book. For the notion of a cleansing—to reveal what is natural beneath the grime of the subnatural—can complement in a remarkable way the understanding of miracles that we have already explored from a more philosophical perspective, providing a new way of looking at the "higher" laws of nature that I have posited.

My understanding of miracles in terms of this complementarity does not, it must be said, represent any standard exposition of the Eastern tradition, which is rarely as systematic as this. Nevertheless, an implicit anticipation of such an approach is to be found in the writings of many Eastern theologians. Moreover, to speak in terms of this kind of complementarity is, as we have seen, consonant with the views of a number of Western theologians. In particular, as I noted in chapter 5, this view is comparable to those of Robert John Russell, who speaks of the possibility of a unique event of religious significance being a "first instantiation of a new law of nature," and of John Polkinghorne, when he speaks of miracles as being analogous to changes of regime in the physical world. These approaches, like my own, hold that there are laws of nature that can, if only under very unusual circumstances, bring into effect what is sometimes referred to as a breaking in of the age to come. The Eastern perspective I have outlined in this chapter adds to this a profound theological interpretation of how these unusual circumstances are to be understood.

As we noted in chapter 5, however, to speak of a "breaking in" is, in this context, somewhat misleading, since what is envisaged in my model is not a breaking in of something that comes from "outside." Rather, I envisage something that the Eastern Christian tradition has often stressed: a "breaking out" of something that is always present in the world, albeit in a way that is usually hidden from us. As I have argued—and as Russell's and Polkinghorne's thinking indicates—we can speak of this in terms of the coming into operation of laws of nature usually inoperative because the conditions for their operation have not been met. As we have seen in this chapter, however, we can also speak of it in terms more familiar to the Eastern Christian tradition: as a transformation from the subnatural to the natural, brought about by an unveiling of God's hitherto veiled presence in all created things.

Because my model envisages the possibility of speaking in terms of "laws of nature" to describe all events, it is clearly, in one sense, an example of a strong theistic naturalism. However, we must be very careful how we use this term, since our culture tends to think of naturalism in terms of what can be uncovered—at least in principle—by the scientific method. As we saw in chapter 5, however, several aspects of the model indicate that the scientific method will not straightforwardly apply to the investigation of miraculous events, and we have now, in the context of the Eastern tradition, seen more clearly why this is the case. Quite simply, the occurrence of these events involves something that cannot be replicated under laboratory conditions: the faithful response to God of those who recognize God as their creator and redeemer.

Therefore, the term *naturalistic* applies to this model only if we take it to have a meaning related not to epistemology (to that which may be known in a particular way) but to ontology (to the way the world is). The naturalism I have outlined is clearly of this latter kind, even if its picture of "the way the world is" is far more complex than the kind of picture favored by most who think of themselves as naturalists. As we have seen, however, what they call naturalism is—both for the Eastern patristic understanding and for my own—in fact no more than subnaturalism. It may be that a true naturalism cannot ignore what a scientific subnaturalism tells us, because the natural and subnatural worlds are intimately related. Equally, however, a true naturalism must go well beyond what a subnaturalism of this kind is able to say. It must be informed not only by the sciences, as that term is usually understood, but also, and even mainly, by what was once called the queen of the sciences: theology. Only in the context of what has been revealed to us by God can the universe in which we live be fully understood.

13

The Word Made Flesh

We have now come to the point at which, through an examination of precisely what it means to speak of the incarnation, we can develop the kind of pansacramental naturalism I have advocated into what I call an *incarnational naturalism*. This examination must begin with the main biblical source of the doctrine: the prologue of the Fourth Gospel (John 1:1–14).

As we have seen, this prologue, when read attentively, clearly indicates that only in the context of a particular understanding of the whole creation can the significance of the historical Jesus be comprehended. For according to this passage, it was not only through the Word—made flesh in Christ—that "all things came to be," but in addition, "All that came to be was alive with his life" (John 1:1–4). Therefore, not only does the Fourth Gospel posit an intrinsic link between the creation of the cosmos and what occurred historically in the person of Jesus, but there is also a clear indication that, in some sense, the Word that "came into" the world in the person of Jesus had not previously been absent from it. As Stephen Need has noted, the incarnation in Jesus "is not the sudden arrival of an otherwise absent Logos, but rather the completion of a process already begun in God's act of creation."[1]

We can understand this more clearly, Need observes, when we recognize the way in which the Hebrew concepts of word (*dabar*) and wisdom (*hokma*) had already been brought together before the time of Jesus. In particular, he notes, many Jews had, by the first century, adopted the sort of Hellenistically influenced philosophical framework associated with writers like Philo of Alexandria, in which the Greek concept of word (*logos*) was important. They had, through this framework, "combined all the meanings of *logos* with all the meanings of *dabar*." Because of this, while there may have been no systematic doctrine in their thinking about the divine *Logos*, they saw it, as Philo did, as "a cosmological phenomenon connected with wisdom . . . associated with God's activity in creation, and with God's image in human beings."[2]

The *Logos* concept used in the Fourth Gospel was, therefore, already well understood both by that gospel's author and by his intended audience. It was used to indicate "a pre-existent aspect of God's being; eternal wisdom; God's instrument of creation and vehicle of revelation; and a rational element permeating and upholding the universe. It was God's dynamic way of communicating his nature and his will to his people and was by definition active in creation."[3] The writer of the gospel—quite possibly using a *Logos* hymn already in use[4]—simply drew on all these meanings and claimed, in addition, that this *Logos* had been made flesh in Jesus of Nazareth.

The use of the term *Logos* by the Christians of the early centuries was not, however, confined to the Fourth Gospel and to direct commentary on it. Later Greek-speaking theologians would, essentially, develop this biblical *Logos* concept in a systematic way. Some of them, for example, attempted to do this in terms of the idea of the world being, in some sense, God's body (a concept that has, as we shall note presently, also been attractive to theologians of our own day). In this respect, many of them suggested a view of the *Logos* that was essentially a modification of the Platonic notion of the World-Soul.

The dangers of this particular route were considerable, however, and some of these attempts—those of the Gnostics and of Arius, for example—clearly failed to develop a sufficiently distinctive usage of Platonism. Because Plato had made the world an inferior imitation of the world of "ideas," they could, and did, fall into the trap of envisaging this World-Soul as a distinct being, inferior to God. However, the more orthodox theologians of Alexandria were able to resolve this problem by developing the Judaic picture of the goodness of God's creation through the concept of God's direct immanence in the world. This enabled them, from Justin onward, to develop these ideas in a more distinctively Christian way, with the *Logos* concept still being used in a way analogous to that in which the Platonists had used the concept of the World-Soul, but without a Platonist denigration of the empirical world.[5]

It was, perhaps, in the thought of Athanasius that this particular picture was brought to its greatest development. For Athanasius, the *Logos* was the source of coherence and order in the universe, and his theory of creation through "condescension" enabled him to avoid the potential dangers that others had failed to avoid. It allowed him to see the eternally begotten Word as effectively uniting with finite things as World-Soul, by accommodating God's nature to their capacities.

This particular concept was, however, never fully taken up by later theologians (even though, in the view of Eugene TeSelle,[6] the condescension model renders spurious the caution that led to this, and may even be seen as a resource for the concept's revival). Rather, it was through the work of Maximos the Confessor that aspects of this Alexandrian understanding were combined with insights from the Cappadocian fathers and others, to produce a panentheistic understanding that remains definitive in Eastern Christian theology to this day.[7] This is the understanding that adopts what we might call a *Logos* cosmology: a view of the entire created order and of its redemption based, in part at least, on the nuances of the Greek term *logos*.

As Andrew Louth has expressed it, to say in Greek "that the universe is created by the *Logos* entails that the universe has a meaning, both as a whole and in each of its parts. That 'meaning' is *logos*: everything that exists has its own *logos*, and that *logos* is derived from God the *Logos*. To have meaning, *logos*, is to participate in the *Logos* of God." Behind this, says Louth, "lurks the Platonic idea, that everything that exists" does so "by participating in its form, or idea, which is characterized by its definition; the Greek for definition (in this sense) is, again, *logos*. These Platonic forms, or *logoi*, to call them by what defines them, are eternal."[8]

However, Louth stresses, between Plato and Maximos "much water had flowed down the history of ideas, and for Maximos, because the world had been created by God through his *Logos*, it can no longer be regarded as a pale reflection of eternal reality, as in Plato's world."[9] For Maximos, this understanding led to an expansion of what Philip Sherrard has called the Greek patristic concept of "the universality of the Incarnation," in which the *Logos* is seen as incorporating itself "not in the body of a single human being alone but in the totality of human nature, in mankind as a whole, in creation as a whole."[10]

Maximos saw the *Logos* of God as being manifested not only in the person of Jesus, but also in the words (*logoi*) of all prophetic utterance, and in the *logoi*—in the sense of underlying principles—of all created things. Christ the Creator *Logos* was understood by Maximos to have implanted in every created thing, at the moment of its creation, a characteristic *logos* (a "thought" or "word") that manifests God's intention for the thing and constitutes its inner essence, making it distinctively itself and drawing it toward God.

These *logoi*, though inhering in each created thing, are not themselves created. They are, for Maximos, nothing other than God's presence in each thing: a manifestation of the *Logos* itself. Because of this understanding, Maximos's thought presupposes, as Lars Thunberg has put it, "almost a gradual incarnation."[11] The person of Jesus represents, in this perspective, not (as in so much Western theology) something essentially alien to the natural world, but rather the coming to fullness of something present in it from the beginning. In this way, the theology of Maximos accurately expresses the insight that, as we have seen, is intrinsic to the prologue of the Fourth Gospel: that "the incarnation in Jesus is not the sudden arrival of an otherwise absent Logos, but rather the completion of a process already begun in God's act of creation."[12]

Why, we must ask, is this notion likely to seem strange to many in the Christian West? The answer seems to lie partly in the way in which Western thinking about the incarnation has tended to take its bearings, not primarily from the biblical or patristic roots of the concept, but instead from Anselm's classic question: *Cur Deus homo?* (Why did God become man?) Although answers other than Anselm's own have often been preferred, most of them have focused, as his did, entirely on human sin and its remedy.[13] In doing so, they have either ignored or forgotten the strand of thinking common in the East—and explicit in the works of people like Maximos and Isaac the Syrian—which holds that "even if man had never fallen, God in his love for humanity would still have become man."[14]

The roots of this divergence in understanding between East and West go much further back than the time of Anselm, however. For while the Greek-speaking Christians of the East were clarifying the Fourth Gospel's *Logos* concept during the church's early centuries, the Christians of the West were simply becoming less and less familiar with the Greek language in which the Gospels had been written. By the time of Maximos, the Fourth Gospel was read in the West, even by scholars, mostly in Latin translation, and *Verbum* (the nearest Latin equivalent of *Logos*) simply did not have the range of nuances of the Greek term.

To say that Western theologians became completely unaware of these nuances would, of course, be an overstatement. But it would be true to say that they only rarely assimilated fully the understanding of these nuances that was almost instinctive among the theologians of the East. In particular, while Western theologians were not entirely unaware of the concept of participation that informed the development of classical christological doctrine, and continued for centuries to use this and related ideas in their contemplative tradition, they never, as Philip Sherrard puts it, formulated these ideas "with such clarity or completeness and . . . their full significance was from the start vitiated by St. Augustine's teaching on sin and free will which became so inextricably intertwined with the western Christian tradition."[15]

Moreover, Sherrard observes, this early divergence was, from the thirteenth century onward, sharply amplified by the growing dominance of Aristotelian categories in the mainstream thinking of Western Christendom. To all intents and purposes, he says—and here he echoes the views of commentators as diverse as C. G. Jung and Paul Tillich[16]—"The Platonic elements which had served the earlier theologians as a vehicle for expressing an understanding of man confirmed through a life of prayer and contemplation were replaced by or codified in accordance with Aristotelian categories of a purely abstract and theoretical nature."[17]

The result, according to Sherrard, was this:

> Where Christology is concerned . . . the union of the divine and the human natures of Christ could no longer be conceived as the union of two substances, each preserving its own integral identity in and through the union: the whole idea of the *perichoresis* [mutual indwelling] as envisaged by the Greek fathers was precluded, as also was the idea that Christ can be the ultimate ground or subject of each single person, who is thereby "deified." The incarnation could be envisaged as something that occurred only in the unique case of the historical figure of Jesus, and not as something that involves human nature as a whole and so something in which every individual participates.[18]

In this way, he says, the "magnificent scope of the Logos doctrine with its whole 'cosmic' dimension—the idea of God incarnate in all human and created existence—which from the time of the Alexandrians and Cappadocians down to the present day has been one of the major themes of Orthodox Christian theology, was tacitly but radically constricted in Western thinking."[19]

As we have seen, this Eastern understanding informs Sherrard's positive attitude to the other faiths of the world, which is very similar to that which arises from my own psychological-referential model of revelatory experience. This attitude is not a modern one, however, for as we have noted, there was often a sympathetic attitude toward pre-Christian religious frameworks of a pagan kind in the patristic period. As Sherrard observes, this was based precisely on an understanding of the doctrine of the incarnation, in which the *Logos* was seen as incorporating itself, "not in the body of a single human being alone but in the totality of human nature, in mankind as a whole, in creation as a whole."[20]

Only later, Sherrard observes, was this early patristic view of the incarnation overlaid by a narrower, purely historical one of the kind that led to the exclusivism that has since characterized conservative Christian theology. He insists, however, that this narrowing is illegitimate, since what he calls the economy of the divine *Logos* "cannot be reduced to His manifestation in the figure of the historical Jesus."[21] What passes for a traditional understanding of the incarnation and of other faiths must, he insists, "be replaced by a theology that affirms the positive attitude implicit in the writings of Justin Martyr, Clement of Alexandria, Origen, the Cappadocians, St. Maximos the Confessor and many others."[22]

What are we to make of these diverging understandings of East and West? Have both moved equally far from the meaning of the Fourth Gospel, or has one side retained, at least broadly, its fundamental insights? Certainly, there can be legitimate argument about this question. In at least one respect, however, it seems to me that the Eastern approach is markedly superior. As we have seen, current exegesis of the Fourth Gospel emphasizes that the incarnation in Jesus was not the sudden arrival of an otherwise absent *Logos*, but rather "the completion of a process already begun in God's act of creation."[23] And whatever its shortcomings may be in terms of an overreliance on Platonic categories and assumptions, the Eastern understanding of the incarnation has retained this perception with a clarity that is rarely found in the West.

Indeed, when some Western theologians have attempted to develop their christological understanding along these essentially biblical and patristic lines, the reaction among their colleagues has often been extremely negative. For example, when Arthur Peacocke once suggested that what is new and unique in the person of Jesus might be understood in part by analogy with the evolutionary emergence of new species or levels of complexity in the created order,[24] one of the main objections that was raised was expressed in very interesting terms. The incarnation, it was insisted, should not be thought of as "the culmination of God's continuing action in creation."[25] This reaction typifies the narrowness of the way in which the doctrine of the incarnation is commonly understood in the West. For although there are clear differences between Maximos's concept of "almost a gradual incarnation" and what Peacocke's critics sometimes describe as his "evolutionary" Christology, these do not lie in the sort of separation of creation and redemption that Western critics tend to adopt at a quasi-instinctive level.

In the context of our present exploration, however, the arguments for and against any particular christological model are less important than recognition of the wider presuppositions that lie behind the Eastern patristic understanding of the incarnation. In particular, it is important to appreciate that this strand of thought includes an understanding of "nature" that is quite different from that assumed in most Western understandings of divine action. These latter, whether of the interventionist or of the causal-joint kind, assume that there is a "natural" or "normal" state in which the cosmos exists, to which something must be added in acts of divine grace. In the Eastern patristic tradition, by contrast, this assumption is not held in anything like the same way. Instead, this tradition posits an essentially dynamic universe in which, as Vladimir Lossky has put it, there is "no natural or normal state, since grace is implied in the act of creation itself. The world, created in order that it might be deified, is dynamic, tending always towards its final end."[26]

Thus, for this tradition, the presence of the divine *Logos* in all created things provides, as we shall see further presently, a kind of teleological dynamism that draws them toward their intended final end, and this dynamism represents a mode of divine action that is neither the "special" nor the "general" one of Western thinking. The universe is, for the Eastern Christian tradition, neither a benignly designed machine nor something that needs to be acted on "from the outside."

An interesting outcome of this view is the way in which the sacraments of the church are understood. In the Eastern tradition, they are, as we have seen, not simply an outward and visible sign of an inward and spiritual grace. They are also, and even primarily, what Alexander Schmemann calls "a revelation of the genuine *nature* of creation."[27] But even if Western Christian thinking usually manifests very little of this sense of the cosmic significance of the sacraments, we must not think that this characteristic Eastern stress is entirely without its echoes in the West. Such echoes are manifested, for example, in something we have already noted: the perception, among some Western Christians, of what Arthur Peacocke calls "a real convergence between the implications of the scientific perspective on the capabilities of matter and the sacramental view of matter which Christians have adopted."[28] When the cosmos is examined through the eyes of the scientist, he seems to suggest, there are clear echoes of those strands of Christian thinking that see the sacraments in terms of the transparency of created things to God's purpose.

The fact that Peacocke can speak in this way not only indicates that the Eastern understanding of the sacraments is at least implicit in important strands of Western thinking. It also implies that any Western framework based on what I have called a pansacramentalism can be enriched and deepened by Eastern incarnational insights. In particular, it suggests the possibility that the pansacramental naturalism I have advocated may be incorporated into what we might call an *incarnational naturalism*, in which Eastern incarnational insights are taken fully into account.

This enriching of the perspectives of one side of the Christian world by the other will not necessarily be one-way traffic, however. We must also consider the possibility that the Eastern notion of the incarnation can itself be enriched by the kind of scientifically informed Western pansacramentalism we have noted.

That this possibility exists is, I believe, indicated by two interconnected aspects of the thinking of Maximos the Confessor. The first of these is that the *logoi* of created things, of which Maximos speaks, are considered by him to be "inviolable."[29] While obscured by the Fall, they are neither distorted nor destroyed, so their role of drawing created things toward their ultimate intended end is, to some degree, still operative. (Here the subtle relationship we have noted in Eastern Christian thought between the natural world and the empirical, subnatural one is important.) The second aspect of Maximos's thinking that is relevant here is the fact that this notion of inviolability does not imply that created things must be thought of in static categories. Rather, while Maximos assumes, with all his contemporaries, that natures are fixed, his thought is, as Andrew Louth notes, still dynamic enough to be implicitly open "to the idea of evolution . . . as a way of expressing God's providence."[30] Put together, it seems to me, these aspects of Maximos's thinking suggest that Louth is correct to suggest that Maximos's cosmic vision can "be re-thought in terms of current science."[31] In particular, I would argue, this can be done in terms of the sort of Western pansacramentalism that is based on evolutionary insights.

Therefore, the Eastern concept of an intrinsically "dynamic" universe does not merely have an immediate and strong resonance for those who already acknowledge the validity of the modern scientific perception of the universe's evolutionary development. In addition, elements within it suggest that Eastern theologians should, by virtue of their own tradition, be receptive to scientific ideas. Because of this, not only can the kind of pansacramentalism that already exists in the Western side of the Christian world be extended in terms of the incarnationally focused notion of the sacramentality of creation that is common in the East, but also the Eastern notion can be extended in terms of the Western pansacramentalism, and at least some degree of convergence will inevitably result. Indeed, I shall argue, this convergence, already implicit in important strands of Eastern and Western thought, can give rise to a unified model. As we shall see, however, this can happen only if both sides are willing to modify their existing perspectives in ways suggested by philosophical and scientific insights into our world.

14

Ecological and Feminist Perspectives

The kind of incarnational approach I have developed, while important in its application to some of the more abstruse questions of the academic theologian, is not limited to those applications. For, as we have already seen, important aspects of it have a direct bearing on both our general spirituality and our ethical thinking and action. So far, however, one aspect of this more practical application has been mentioned only in passing. This is the way in which allowing an incarnational pansacramentalism to percolate to every level of our being results in a sense that we need to revere and care for the creation. Any previous sense that the rest of creation exists simply to serve human needs and has no intrinsic value simply evaporates, and we begin to see as extraordinary the fact that our Christian forebears so often interpreted the "dominion" notion from the book of Genesis in this way.[1]

It would, however, be foolish to claim that the form of pansacramentalism I have advocated represents the only strand of Christian theology that is likely to have this effect. For a generation at least, a steady stream of studies have argued not only that the biblical exegesis that has given rise to the common notion of "dominion" has been mistaken, but also that at the heart of the Christian tradition there exists, either potentially or actually, something akin to the reverence for the world that characterizes what is sometimes called the "deep ecology" movement. Many of these studies have been written from a specifically ecological or feminist perspective, and one of these, Sallie McFague's book *The Body of God*,[2] is particularly relevant to our exploration here, since it focuses on defending the metaphor of the world as God's body, which clearly has strong panentheistic overtones of the kind found in my own model.

By saying that my model can be interpreted in a creative and nuanced way through McFague's kind of panentheism, I do not mean to imply that this interpretation will be free of creative tensions. Despite the profound parallels between aspects of her model and mine, I believe that several aspects of her work require modification. Many of these have their roots in the way in which McFague—like many commentators working within an ecological or feminist framework—has a somewhat schematic grasp of the history of Christian ideas.

For example, when McFague speaks about what she calls the "organic" model she advocates, she asserts that the oldest Christian form of this model—based on passages from Ephesians and Colossians dealing with the cosmic Christ—was at an early stage in Christian history "narrowed to human beings and especially those who acknowledged Christ as the deity. It lost its cosmic reach, the inclusion of the natural world and all human beings."[3] While she does recognize that some early theologians spoke of the world "as a body filled with and ordered by the Logos in a manner similar to the Platonic World-Soul," she asserts that this kind of "intimacy between God and matter came to an abrupt end when, in the Nicene faith, the Logos became identified exclusively with the second person of the trinity, with the transcendent God."[4]

As we have seen, however, while the World-Soul concept did fall into disuse in the post-Nicene period, the kind of "intimacy between God and matter" it had attempted to delineate was by no means forgotten or abandoned. Only in certain strands of Christian thinking—and even in these, often much later than the immediate post-Nicene centuries—did it become true that there arose, as McFague puts it, "a profound difference between the . . . early Christian notion . . . of the cosmos animated by the . . . Logos of God and the church as the body of Christ."[5] For example, when in our own generation, Olivier Clément speaks of the "human (and cosmic) material united in Christ with the godhead,"[6] or when Panayiotis Nellas speaks of "Christ . . . the Logos together with His body, the transformed world,"[7] they are speaking from within a tradition that stands in unbroken continuity with that within which the Nicene faith was formulated.

It can, of course, be argued that the contrast McFague stresses is now an intrinsic part of the intellectual and spiritual heritage of most who call themselves Christians, and that her simplification of history is therefore not crucial to an assessment of her more positive suggestions. This argument is, I believe, at least partially valid. For instance, her recommendations about how the effects of this dubious inheritance can be overcome are certainly far from worthless; in particular, there is considerable merit in her sense that an importance resource for this lies in the scientific account of the universe's development and in an ecological sensibility.

Her feminist and ecological awareness of oppression is also, I believe, important, since it makes her cognizant of some of the problems inherent in certain understandings of how God has created the world through the evolutionary process. In particular, she rightly notes that one of the major shortcomings of "creation spirituality," as advocated by Matthew Fox and Thomas Berry, lies in the way in which its celebration of cosmic evolutionary perspectives is linked to insensitivity to the suffering in the world. For example, she sees Berry's perception of the world as "a single gorgeous celebratory event" as being helpful insofar as it reflects a spirituality that "redresses centuries of an increasingly narrow focus of divine concern on human beings, especially Christian human beings."[8] However, she rightly goes on, this phrase also indicates creation spirituality's greatest weakness: too often, those who speak in these terms lack "a sense of the awful oppression that is part and parcel of the awesome mystery and splendour. The universe has not been for most species, and certainly not for most individuals within the various species, a 'gorgeous celebratory event.' It has been a story of struggle, loss and often early death."[9]

This awareness of the problem of natural evil is, as we have seen, something that I believe to be important, and in this respect, my own critique of creation spirituality is similar to McFague's. In her thinking, however, creation spirituality's lack of such awareness is inextricably linked to its claim—often reflecting the influence of Pierre Teilhard de Chardin—that an optimistic view of the evolutionary story should be adopted. However, she does not simply criticize the "theological Stalinism"[10] of de Chardin's views, in which the suffering in the world is effectively ignored on the grounds that the end justifies the means. (A criticism on this basis would, in my view, have been entirely justified.) She also criticizes the very notion of an intended directionality within the evolutionary process, which seems to her both to smack of the God-of-the-gaps approach and to posit a God "with a narrow job description by Hebrew or Christian standards."[11]

Here, perhaps, more than anywhere else in her study, McFague's incomplete grasp of the Christian tradition most seriously affects her argument. For the kind of tradition-oriented view that I have outlined in this book—and to which, as we have seen, these criticisms of the notion of creation through evolution do not apply—does not seem to occur to her even as a possibility.

Even if we can see serious shortcomings in McFague's approach, we can also surely recognize the way in which her metaphor of the world as the body of God can be extremely creative in relation to the perspectives I have outlined. Her model might pose dangers if taken too literally so that a one-to-one correspondence between specific aspects of "God's body" and of created bodies were attempted. Here, however, her nuanced (and perhaps even unduly instrumentalist) understanding of the use of models and metaphors keeps her from succumbing to these dangers and allows her to move us subtly and often impressively toward a model that is, like the one I have advocated, panentheistic,[12] christological,[13] and even (though she does not use the word) pansacramentalist.[14] Despite the shortcomings I have noted, McFague provides us, in my judgment, with important considerations that complement and illuminate aspects of the model I have set out here.

In a rather different way, something of what I have said can also be illuminated by another attempt at an ecologically sensitive theology: that of Mark Wallace's book *Finding God in the Singing River*. In many respects, it must be said, Wallace's approach is far less nuanced than McFague's. In particular, if McFague has an important awareness of the suffering in the world, Wallace (judging by what he says) sees this suffering in only a limited way. Similarly, if she reads the Christian tradition through distorting spectacles, he, for all intents and purposes, simply ignores it and manifests the kind of approach that is now far from uncommon in Western academic theology, in which little or nothing of the Christian thought that occurred between the biblical writers and the postmodernists is considered relevant to current concerns.

Despite all this, Wallace's approach offers, in my judgment, something that deserves our serious attention. Not only does he insist on the need to develop the Christian notion of incarnation in a direction similar to that which I have advocated, so that it affirms the concept of God as "continually enfleshed with the natural world as we know it,"[15] but he also (and, in my view, extremely perceptively) sees the relevance of contemporary neopaganism to our understanding of how this can be done in the Western Christian context.

Wallace does not here underestimate the strength of the antipathy toward neopaganism among Christians of the present day, and of the tendency (as in the work of Carl Braaten) to define neopaganism as "a catchall for everything opposed to Christianity."[16] However, he notes, the reality of neopaganism is only rarely understood by Christians, and much in its practice, and even in its beliefs, may be seen as consonant with a "green" Christianity and as a resource for its development. Many Christians, he rightly says, when faced with the beliefs and practices of contemporary neopagans, simply "do not recognize the origins and ongoing vitality of their religion in biblical and Pagan teachings that the earth is holy and that all things are filled with the Spirit—that all things carry an 'innate divine seed' as . . . Braaten (disparagingly) puts it."[17]

It is primarily in the last part of this statement that we can, at a theoretical level, see the relevance of Wallace's perspective to my own. For as he points out, Braaten—a Protestant theologian of a fairly typical kind—actually goes so far as to define the prime anti-Christian aspect of neopaganism in terms of "variations of the ancient belief of pre-Christian mystery religions that a divine spark or seed is innate in the individual human soul."[18] As we have seen, however, this ancient belief was not limited to the mystery religions of the late antique world, but was actually a central element of the thinking of the early Christian centuries, which took its bearings from the *Logos* concept, which had arisen from pagan philosophy as well as from the Hebraic notion of divine Wisdom.

The point here is that, for people like Braaten—as for Karl Barth and for many Protestant theologians of the past century—the Christian revelation is essentially unconnected to the world and the human condition as we experience them. For a truly incarnational theology, however, that revelation represents the coming to fullness of what has existed from the beginning. In this perspective, the very notion of incarnation leads us to appreciate the way in which (as Louis Bouyer once put it of another aspect of the parallels between early Christianity and the mystery religions) "the divine reveals itself in the transformation it effects on what is human."[19]

The importance of Wallace's focus on neopaganism lies not only in these considerations, which merely indicate the legitimacy of this focus at a theoretical level, but also, and perhaps primarily, in his sense that neopaganism represents both a major element in the somewhat unfocused spirituality of many people in the present-day West and an important resource for the development of a "green" Christianity.

At one level, the neopagan contribution to this development is likely to occur at the level of general spirituality and liturgical observance. In the Western churches especially, there is a real lack of liturgical focus on the cosmic aspect of the Christian faith, and the adoption and adaptation of certain kinds of neopagan practice might prove a helpful corrective to this. (A good example might be the kind of daily ritual that Wallace himself observes, based on a strand of Native American practice.[20]) At another level, however, the fruits of Wallace's approach may well be more explicitly theological. In particular, his sense of the way in which the scriptural texts "figure the Spirit as a creaturely life form interpenetrated by the material world"[21] may have much to teach, not only to those whose focus will be largely on biblical interpretation, but also to those who already accept the sort of patristic *Logos*-focused framework I have advocated.

I say this because it seems to me that one of the problems of using the *Logos* concept as our main way of speaking about God's presence in the world is that it can easily be interpreted in a way that disregards important aspects of the biblical foundation of classical Trinitarian doctrine. To speak of God's relationship to the creation in terms of the *Logos*—identified in classical Christian doctrine with the second person of the Trinity—can easily lead to a lack of proper emphasis on the third person of that Trinity: the Holy Spirit of God, who according to the book of Genesis (1:2) was instrumental in the creation. This can lead to a number of distortions in our theological thinking, not the least of which is something that has been noted by Wallace and by many feminist theologians: that the feminine imagery associated with the biblical notion of the Spirit tends to be lost to general use.

This aspect of Wallace's critique is, however, dealt with in a more thorough and balanced way in other strands of contemporary theological thought than his own. For example, an exemplary combination of theological and scientific reflection—informed by feminist and ecological insights, but not determined by them—is to be found in the work of Celia Deane-Drummond. In particular, her book *Creation through Wisdom*[22] attempts, in an impressive way, to develop a theology of creation that does full justice to the Old Testament figure of Divine Wisdom, *Sophia*, who in the book of Proverbs (8:22) speaks of herself as "the beginning of [God's] works, before all else that he made."

This figure, as a number of feminist theologians have stressed, not only is more unambiguously feminine than is the figure of the Spirit of God, but also was, historically, an important element in the Jewish appropriation of the *Logos* concept used in the Fourth Gospel. Unlike many feminist authors, however, Deane-Drummond does not simply stop at the wonderful biblical imagery of the book of Proverbs as if, in isolation, it proved an important point. Rather, she analyzes it carefully, discussing both the more anthropologically focused aspects of Wisdom to be found throughout the Old Testament's wisdom literature, and also what she calls the "wisdom of the cross" to be found in a number of strands of New Testament thinking.

In combining all these insights and attempting a synthesis, Deane-Drummond sees certain strands of Eastern Christian thinking as important, quoting especially two figures whose significance I have already noted: Gregory of Nyssa and Maximos the Confessor. However, the full flowering of the kind of approach she sees as relevant does not, for her, occur in the thought of the patristic period itself, but in the "sophiological" speculations that were, after their development in nineteenth-century and early-twentieth-century Russia, to prove controversial in the Eastern Orthodox community within which they arose.

Deane-Drummond discusses, in particular, Sergeii Bulgakov's attempt to refute the accusation that his thought is incompatible with the classical Trinitarian thinking of the church. This attempt was based, she explains, on a bid to unite his earlier sophiological speculations more clearly with classical Trinitarian doctrine by linking the Sophia figure to the essence (*ousia*) of God. In this way, she says, Bulgakov saw Sophia as "the 'Eternal Feminine' of the God-head, becoming *both* the internal love of God *and* the connection between the created world and the eternal world of the triune God."[23]

The strong reaction by many Eastern theologians against this thinking is, for Deane-Drummond, questionable. In particular, she argues that the insistence of some on the classical identification of Wisdom with the divine *Logos*, and of others that Wisdom is no more than one of God's "energies," is problematic. "While Bulgakov's Sophiological scheme is perhaps too elaborate," she says, "to reduce Wisdom to just another energy like Justice or Life seems . . . to fly in the face of the biblical tradition of Wisdom as person. Furthermore, to restrict Wisdom to Christology does not do justice to the variety of wisdom texts."[24]

Deane-Drummond's own suggestion here is that Wisdom should be seen primarily as a characteristic of the Holy Spirit:

> It is only in the power of the Holy Spirit that the Logos can become the Wisdom of God. Christ becomes God incarnate through the power of the Holy Spirit, as the story of the annunciation makes clear. It is only through the Holy Spirit that the cross expresses the Wisdom of God. It is only through the dynamic involvement of the Holy Spirit that Christ is raised from the dead in the resurrection. The identification of Wisdom and Logos does not mean that the two are indistinguishable, rather one cannot be thought of without the other.[25]

Basing her reflections not only on this kind of consideration, but also on many others, including the early patristic usage of Irenaeus and others, Deane-Drummond argues for an understanding of creation through Wisdom that is more clearly Trinitarian than is often the case when the *Logos* of God is invoked as the agent of creation. Her approach remains an incarnational one but is expressed in terms of the way in which "the joint action of Word and Wisdom in creation is expressed in the incarnation of the Son." She explains, "In the person of Jesus, the Son becomes both Wisdom incarnate and Word incarnate. This is a unique penetration of the Godhead with the material world. Whereas in the act of creation Word and Wisdom serve together to shape the creative process, in the incarnation of the Son both Word and Wisdom become part of the fabric of creation itself."[26]

This suggestion seems to me to be problematic in much the same way as Bulgakov's own. While Deane-Drummond is, in my view, right in seeing the reaction to Bulgakov's speculations as inadequate, her own use of the Wisdom concept—as if it is logically and grammatically equivalent to the *Logos* concept, yet without identifying the two—leaves her in danger of implicitly positing, as Bulgakov is sometimes accused of doing, a fourth hypostasis to add to the three posited by classical Trinitarian theology.

There does, nevertheless, seem to be a genuine insight in Deane-Drummond's sense that the classical use of the *Logos* concept is in danger of failing do justice to the biblical concept of Divine Wisdom. Her arguments indicate, I believe, that we do need to think more clearly than we usually do about how the Wisdom figure of the Old Testament should be related both to the *Logos* concept and to the work of the Holy Spirit as seen in the New Testament. In particular, if we are to use the evident parallels between the biblical descriptions of *Sophia* and *Logos* to equate the two, we must begin to think more seriously than we usually have about whether the historical incarnation of the *Logos* in a male figure, Jesus of Nazareth, has led us to think of the *Logos* itself in terms that preclude its feminine aspect.

For example, we must ask ourselves why we can all too easily translate the patristic term *logikos* (meaning that which pertains to the *Logos*) as "logical," when the Greek term in fact has many more complex and "feminine" nuances of the sort that the figure of Divine Wisdom symbolizes.[27] Does the answer to this question lie at least partly in the way we have not only ignored the feminine imagery associated with the Holy Spirit but also tended to put less emphasis than is appropriate on the way in which whatever is done "through" the Son is also, in early Christian thought, done "by" that Spirit?

Here one of the problems, at least among Eastern Orthodox theologians, might lie in the tendency to put little emphasis on the kind of position articulated by Gregory Palamas, when he explicitly quotes the Wisdom passages in Proverbs to underline the way in which the Eastern tradition, while insisting that the Spirit proceeds from the Father alone, can still affirm that "the Spirit belongs to the Son, who receives Him from the Father as the Spirit of Truth, Wisdom and Logos," so that the "pre-eternal rejoicing of the Father and the Son [in one another] is the Holy Spirit who . . . is common to both."[28] Certainly, among more recent Eastern theologians, the tendency has been to put such emphasis on the Spirit as a separate hypostasis of the Trinity that this (now predominantly Western) notion of the Spirit as the mutual love of Father and Son is underemphasized or even ignored altogether.

The nuances of Trinitarian theology are, however, extremely subtle, and I do not here want to base my argument too much upon them. Nevertheless, it does seem to me that when factors of the kind I have mentioned are properly taken into account, it becomes possible to use the classical *Logos* understanding in a more nuanced way than has often been the case. Without adopting Deane-Drummond's model, we can still achieve the kind of balance she claims for it, in which we can "image God according to feminine metaphors," not in terms of replacing male images with female ones, but "in a complementary manner."[29]

This requires, I believe, not a model of the sort that Deane-Drummond advocates, but something much simpler: a recognition that the divine *Logos*, because it is manifested as Wisdom, is not limited to the "masculine," rational aspects of God with which the *Logos* term has often been exclusively associated. We need to see this limitation as being untrue to the biblical insights on which our Trinitarian theology is based and develop the kind of sensibility that, in the late medieval period, characterized the thought of Julian of Norwich. ("In our Mother Christ," Julian wrote, "we profit and increase, and in mercy he reforms and restores us, and by the virtue of his Passion and his death and Uprising, ones us to his substance. Thus works our Mother in mercy to all his children who are pliant and obedient."[30])

The divine *Logos*, we must recognize, is manifested not only in divine rationality and in the male person of Jesus, but also in all those elements of divine action and love that we apprehend primarily in a "feminine," intuitive manner and often associate with the Holy Spirit. The biblical images of God's relationship to the world clearly manifest this complementarity, and ultimately, I believe, our *Logos* language must do so, too.

15

A New Understanding

We have reached the point at which we can ask whether all the perspectives I have outlined can be used to develop a single model of God's action in the world. In this chapter, I shall suggest that what I have said introduces the possibility of what we might call a neo-Byzantine model, which, while deeply traditional, is little less than revolutionary in its ramifications. To put this model in its full context, let us recap some of the points I have made.

Briefly stated, the problem of divine action has, in recent thought, been whether and how God can affect the workings of a world characterized (usually at least) by obedience to "laws of nature." Essentially, two kinds of answers have been offered.

One of these has relied on a traditional conceptual scheme, which speaks of a "special" mode of God's action that is analogous, in many respects, to that of any other personal agent. In this understanding, a clear distinction is made between general providence, which arises straightforwardly from what the cosmos will do "on its own," and what will happen if God chooses to perform some special providential action by interfering with its normal workings.

The other main kind of response to the problem of divine action is conceptually different from this interference model (though it is sometimes combined with aspects of it). Here the focus is on "natural" causes as secondary ones, which are to be distinguished from God's will, which is the primary cause of each event that occurs in the world. As this scheme is usually expressed, however, it has the problem that it can give rise to outlooks as varied as a traditionalist neo-Thomism and an essentially deistic kind of naturalism. This variety is possible because the model, in itself, has no specific answers to the questions of how God's will is brought about through natural causes and what the scope of this mode of divine action is.

In all that I have said about how a strong theistic naturalism may be expanded, I have, though avoiding the vocabulary of primary and secondary causes, outlined an understanding of divine action that can in certain respects be seen as an attempt to refine this second kind of model. A significant factor in the development of this understanding has been what I have called the pansacramentalism inherent in aspects of the current dialogue between theology and the natural sciences. This has led me to dub this initial approach a "pansacramental naturalism," and I have drawn attention to aspects of the Eastern Christian tradition that I believe might give this approach a firm rooting in traditional Christian thinking about the incarnation. In this way, I have suggested, a pansacramental naturalism can be expanded into what we might call an "incarnational naturalism." I now wish to present my argument that this naturalism can itself be expanded—through incorporation into what might be described as a neo-Byzantine model of God's presence and action in the world—to offer an essentially new model of divine action, which transcends the concept of naturalism as it is usually understood.

The model I shall develop takes its historical bearings from the strand of Greek patristic thinking that culminated in the work of Maximos the Confessor. In this work, as we have seen, the Fourth Gospel's assertion that the *Logos* (Word) of God "became flesh" (John 1:14) is not understood simply as a statement about a historical event. Rather, Maximos develops his understanding through a subtle and profound perception of how everything was, in the beginning, created through this *Logos* (John 1:1–4). By molding the philosophical categories available to him to the realities of the Christian revelation as he perceives them, Maximos expresses his faith in terms of the way in which the *Logos* of God is to be perceived, not only in the person of Jesus, but also, in some sense, in the "words" (*logoi*) of all prophetic utterance, and in the "words" (*logoi*) that represent the underlying principles of all created things from the beginning.

In this way, both Maximos and those earlier strands of the Greek patristic tradition that culminate in his work focus on the concept of God's *Logos*, in a way that points, for some, toward a pluralistic expansion of traditional Christian theology. The incarnation in Jesus is not seen simply in terms of a historical event that is to be interpreted, as in so much Western Christian theology, as a supernatural intrusion into the created order. Rather, what occurs in the person of Jesus is, for this strand of thinking, intimately linked both to the whole history of the redemptive process and to the creation itself. Lars Thunberg's comment that Maximos envisages "almost a gradual incarnation"[1] might be imprecise in its terminology, but it points accurately to an important aspect of Maximos's thought. Just as the Fourth Gospel's prologue speaks, not of "the sudden arrival of an otherwise absent Logos," but rather of "the completion of a process already begun in God's act of creation,"[2] so also Maximos uses the *Logos* concept to describe a continuous process from the beginning of the cosmos to the Christ event.

The insight that Maximos expresses in this way is, as we have noted, an explicit manifestation of a more general intuition that is implicit throughout the Eastern Christian tradition: that it is quite wrong to speak of grace as something added as a supernatural gift to "pure nature." Rather, as Vladimir Lossky has rightly noted, this tradition knows nothing of this "pure nature," since it sees grace as being "implied in the act of creation itself." Because of this, as he goes on to note, the cosmos is seen as inherently "dynamic . . . tending always to its final end."[3]

What Lossky hints at here is the way in which, for Byzantine theology, at least some aspects of divine providence arise from within the creation through the intrinsically teleological factors that have been, so to speak, built into its components.[4] In fact, this is particularly clear in the work of Maximos. For him, the *logos* that constitutes the inner reality of each created thing not only is uncreated and as such a manifestation of the divine *Logos* itself, but also is, as Kallistos Ware puts it, "God's intention for that thing, its inner essence, that which makes it distinctively itself and at the same time draws it toward the divine realm."[5]

Thus, for Maximos and for the strand of the Greek patristic tradition that culminates in his work, the way in which each created thing has its origin and intended final end in God is intimately linked to the constitutive presence in it of a characteristic *logos*, which is a manifestation, in some sense, of the divine *Logos* itself. That presence not only gives to each created thing the being it has in the temporal world, but also draws it—from within and not through some external, "special" action—toward its eschatological fulfillment. This approach posits, then, a model of the created order that is both teleological and christological. It is a teleological model in the sense that created things are continuously drawn toward their intended final end (though not in a way that subverts human free will and its consequences). It is a christological model in the sense that this teleological dynamism comes about not through some external created "force," but through the inherent presence of God's Word in the innermost essence of each created thing.

At the present time, perhaps, few outside of the Eastern Orthodox tradition are likely to accept the details of Maximos's philosophical articulation of this model. The reasons for this do not, however, preclude consideration of what we might call the general "teleological-christological" character of the vision he articulates. Indeed, the adoption of a teleological-christological model of divine action seems to have several advantages in the context of current debate. Not the least of these is the model's way of envisaging, in its teleological aspect, a mode of divine action that is neither the "special" nor the "general" mode of one strand of Western thinking. For, by allowing us to transcend the need for any distinction between what nature can do "on its own" and what can only be done through some "special" mode of action, a neo-Byzantine model of this sort would allow us to see God's presence and action in the cosmos simply as two sides of the same coin. In this respect, it seems not only to tend toward the sort of model that speaks in terms of primary and secondary causes, but also to give this model a far more definitive theological grounding than it has usually been given.

Even so, the model does present important problems to be solved. One of these is the question of how it relates to the Eastern notion of the "garments of skin," which itself, as we have seen, is related to the problem of natural evil as conceived in the West.

Maximos does not take up this problem directly, however, but only indirectly through his notion that the effect of the Fall on the *logoi* of created things is to obscure rather than destroy them. Does he, perhaps, see this notion of the inviolability of the *logoi* as referring not only to their inmost essence but also to their teleological character? If so, then he might see their teleological effect as being obscured because the leading of created things to their intended ultimate end can, in a fallen world, come about only via a more roundabout (and less easily comprehended) route than would have been the case in an unfallen one, especially when the potential for the sinful use of human free will is taken into account. If this is a correct reading of Maximos's intentions, then divine action in this scheme may still be seen as essentially teleological.

Such an understanding of Maximos's model must, however, remain very tentative, since any invocation of his concept of the inviolability of the *logoi* of created things invokes an aspect of his thought of which, as Andrew Louth has rightly noted, "we can only catch glimpses."[6] It is not Maximos's specific understanding that I want to advocate, however, but the more general teleological-christological vision within which it is set. In this regard, the main problem that arises is not that of natural evil, but of how divine action is related to the workings of a world characterized—usually at least—by obedience to "laws of nature."

Here, I would argue, one of the chief difficulties lies not in our belief that the behavior of the cosmos is characterized by such laws, but in the particular way in which we usually think about them. The best way to explore this issue, in the present context, might be in terms of another approach to divine action that eschews the concept of "special providence." This is the understanding with which our exploration in this book began: that which is sometimes labeled strong theistic naturalism and can be developed in terms of what Willem Drees calls "a scheme of primary and secondary causes, with the transcendent realm giving effectiveness and reality to the laws of nature and the material world governed by them."[7]

An important part of the historical pedigree of this strong theistic naturalism is, as we have seen, the deism of the eighteenth century. The deists did not, of course, express their beliefs in terms of primary and secondary causes, and they tended to see God as more definitively separated from the world than such terms suggest. Their approach was, however, in many respects similar to that adopted by most contemporary advocates of a strong theistic naturalism. In particular, deists saw the scope of divine providence as being extremely limited by a naturalistic approach, and in this, the majority of contemporary advocates of a strong theistic naturalism follow them. Both groups not only eschew the concept of special providence, but in addition hold that what is possible through general providence is extremely limited. Neither group, for example, can see intercessory prayer as having any purpose, other than perhaps for refining the religious sensibilities of those who indulge in it. Their understanding simply precludes the possibility that such prayer can have any sort of practical outcome.

It is important to recognize, however, that this is not philosophically the only option available to the strong theistic naturalist. A naturalistic view, in itself, assumes simply that the cosmos develops according to "fixed instructions" of a lawlike kind, and the possibility that such instructions can bring about subtle and appropriate "responses" to events in the world cannot be precluded in principle. This can be seen through the analogy of human providential action.

For example, as we have already noted, parents can financially support their children through a standing order to a bank. Such an order not only can include instructions about the transfer of money on a regular basis (the equivalent of ordinary general providence), but also can anticipate specific needs. Such an order can, for instance, include instructions of this kind: "If my daughter provides a receipt for repairs to her car, then transfer to her account, over and above her regular payment, the amount necessary to pay for those repairs." An instruction of this sort has the effect of special providence—bringing about action in response to a specific rather than a general need—even though it comes about through a "secondary cause" mechanism of the general-providence kind, and no new action on the part of the prime agent is necessary.

Using this analogy to suggest that general divine providence might be extended to account for events of the sort usually attributed to God's special providence does, of course, have limitations. In particular, it can be objected that humans cannot anticipate all possible needs, and an analogy based on a set of if/then statements can provide neither an elegant model of divine providence nor one based on mechanisms that are conceivable. As we have seen, however, neither of these points is strictly relevant.

On the first issue, we simply need to note that God's wisdom is not limited in the way that human wisdom is. There is therefore no reason to insist that God's "fixed instructions" for the natural world must be inadequate in the way that human instructions must. As to the problem of elegance, it is important to recognize that the analogy is not intended to elucidate the mechanism of divine action, but simply to illustrate an important principle: that "responses" of a providential sort can be the result of a fixed-instruction mechanism. The analogy's mechanism clearly relates more to human limitations than to divine possibilities.

It is, of course, far from easy to guess how God may have set up providential fixed instructions less clumsily than we humans must. As we have seen in chapter 4, however, mechanisms for providential "response" to needs are not entirely beyond conjecture. Moreover, the validity of an expanded theistic naturalism does not depend on the validity of any particular mechanisms that may be suggested. It depends, rather, on acceptance of the general belief that lies behind the search for such mechanisms: that the creation, with its inbuilt "fixed instructions," is far more subtle and complex than our present scientific understanding indicates. Some naturalists might find this idea difficult to accept, but it is not incompatible with naturalism as such.

The point here is that the "laws of nature" that can be provisionally identified are those that can be explored through the scientific methodology. Although this methodology may vary somewhat from discipline to discipline, it relies on the repeatability of observation or experiment and on the discernibility of a particular kind of logic of cause and effect. We need to recognize, however, that we cannot preclude the possibility that the cosmos obeys not only the laws that can be identified in this way, but also other "fixed instructions" that are not straightforwardly susceptible to this investigative methodology.

Indeed, this possibility may even seem likely when we consider the effects of complexity. Practical repeatability and discernible cause and effect are characteristic of only relatively simple systems, which can be effectively isolated from factors that would obscure these characteristics. In addition, as we saw in chapter 1, important issues related to reductionism in the sciences suggest the necessity of positing laws or organizing principles of a kind that are not susceptible to ordinary scientific investigation but can only be inferred from their general effect.[8]

As we have seen, this issue of complexity also has important ramifications for our response to anecdotal evidence of phenomena of the kind labeled "paranormal." It is simplistic to see such phenomena as spurious simply because they are not susceptible to investigation through normal laboratory methods. There is nothing incoherent in believing that such phenomena may occur through processes that are, even though they follow lawlike patterns, in practice impossible to replicate in a straightforward manner. The failure of laboratory methods may simply indicate that such phenomena occur only in situations of considerable complexity or extremity.[9] Once this is recognized, the supposed impossibility of paranormal phenomena becomes questionable, and a number of further questions present themselves for consideration. Not the least of these is what weight we should give to the anecdotal evidence for such phenomena, which (in the religious context in particular) we may judge to be considerable.

The point of these considerations is that when deistic naturalists deny the possibility of the kinds of events usually ascribed to special providence, they do so in the context of an inadequate philosophical argument. As we have seen, there is no fundamental reason to insist that such events cannot be ascribed to the regularities of the natural world that have been "built into" that world from the beginning by its creator. A strong theistic naturalism can, in principle, be constructed in such a way that the scope of divine action is not limited in the way that the deists assumed.

Of course, this does not in itself mean that a strong theistic naturalism is theologically acceptable. For example, one of the main objections sometimes voiced to it is that a naturalism of this kind—whatever the scope of divine action it allows—still envisages the essentially "absent" God of the deistic model. And certainly, if one accepts the separation of God from the world that has characterized most Western philosophical theology, this objection will seem valid to many.

In the context of this argument, we must, however, remember that there has been something of a reaction against this concept of separation in recent years. As we have noted, panentheism—the notion that the cosmos is to be seen as being in some sense "in God"—is finding increasing favor from a number of different perspectives.[10] And if, for whatever reason, we do adopt a panentheistic position, then the argument about an "absent" God immediately fails, for if the cosmos is within God's self, then God can hardly be said to be absent from it. (In this sense, it is ironic that a panentheistic understanding is advocated by many, including Arthur Peacocke and Philip Clayton,[11] who defend a causal-joint understanding of God's providential action. For by taking up a panentheistic position, they undermine what might otherwise have been the main argument against the sort of strong theistic naturalism they oppose.) Panentheism makes quite explicit what is, according to Willem Drees, already implicit in a strong theistic naturalism expressed in terms of primary and secondary causes: that God is, for such a naturalism, "the ground of all reality and thus intimately involved in every event—though not as one factor among the natural factors."[12]

If, as this argument suggests, a strong theistic naturalism will be more persuasive if it is expanded in terms of a panentheistic understanding of the relationship between God and the world, then this persuasiveness will be reinforced if such an expansion is based on something more than an ad hoc juxtaposition of the two frameworks. This is, in fact, one of the reasons that a reworked version of the Byzantine "cosmic vision" suggests itself as a candidate for such expansion. For not only, as we have seen, is such a model comparable to a strong theistic naturalism in its rejection of the concept of special providence, but it also arises from a traditional model that, as commentators have noted, constitutes an explicitly panentheistic framework.[13] The possibility of a synthesis of the two frameworks is therefore intriguing. It is also, I would argue, tenable, provided that we can accept a notion that is intrinsic to the Byzantine understanding of God's action but that has hitherto been ignored by advocates of a strong theistic naturalism. This is the notion of teleology.

The notion of teleology, of course, tends to strike a dissonant note in contemporary discussion. This is largely because the development of our understanding of the laws of nature was associated historically with the abandonment of the teleological thinking that had characterized the late medieval thinking of the Christian West. Given this association, many will inevitably ask themselves why they should reconsider a notion that constituted, half a millennium ago, one of the chief barriers to human intellectual progress.

Those who dismiss teleology in this way tend to forget, however, that the chief impediment to the development of early modern science was not teleology per se but the total philosophical framework then current. (Aspects of early modern physics can still, in fact, be expressed in teleological terms.[14]) More important, they ignore the fact that we can no longer regard scientific understanding, or the philosophical framework within which that understanding is developed, in the same way as could those of the early modern period who rejected teleological thinking. A teleological understanding may be at odds with the picture of the "laws of nature" that is commonly accepted within our culture as a result of the lingering influence of the clockwork model of the cosmos that so influenced the deists. It is, however, in many ways surprisingly consonant with a truly contemporary understanding of the character and content of the natural sciences.

Not only has contemporary science challenged many of the broad philosophical aspects of early modern science that would tend to negate a teleological understanding,[15] but it also has actually evoked questions about teleology in a direct way. In particular, it has indicated that a universe whose development depends on laws of nature and on certain fundamental physical constants need not necessarily be a fruitful one of the sort that ours clearly is. As we have noted, only very particular laws, together with very "finely tuned" values of those physical constants, provide the possibility of a fruitful universe like our own.

This insight has given rise to many arguments related to what is usually called the "anthropic cosmological principle,"[16] and from a theological perspective, these need careful analysis. Some, for example, have seen anthropic considerations as allowing the development of an apologetic approach comparable to that of the natural theology of the past. This has not been widely accepted, however. The consensus view, within the dialogue of science and theology, is that the perception that the universe has been able to "make itself" so fruitfully is not persuasive of, but simply consonant with, the notion of its purposeful creation. In this sense, it is widely agreed that this perception provides the foundation, not for a natural theology akin to that of the past, but for a "theology of nature" in which, for religious believers, scientific perspectives provide valid insights into the way in which God acts in the world's continuing creation.

If we accept, with those who think in this way, that God's creative action should be understood at least largely in naturalistic terms,[17] then we are faced with the question of how we should understand the teleological aspect of this viewpoint. Here, two key points need to be made.

The first is that, in speaking of a teleological factor in this context, we are speaking of something very different from the teleological factor assumed in the Aristotelian thought of the late medieval period. The model I advocate does not compete with the concept of mathematical laws of nature, but focuses on the meaningful outcome of the working of those laws. It envisages what we might call a teleology of complexity: a framework in which we can see significance in the increasing intricacy of the cosmos's structures and in the successive emergent properties to which this intricacy gives rise.[18]

Just as it is possible for Simon Conway Morris to talk about evolutionary convergence in terms of predictable, functional solutions to problems of adaptation —"attractors" analogous to those in chaos theory—so here teleology is understood not as in medieval philosophy, but in similar terms to those he has outlined. The interaction of chance and the laws of nature is such, it would seem, that certain developmental paths are in practice very likely to be followed, and these "attractors" may, in a pansacramental perspective, be explicitly understood as a component of the divine intention. What for Conway Morris is simply a guess about the outcome of a scientific research program is here taken as axiomatic for a theological model: that there is "a deeper fabric in biology in which Darwinian evolution remains central as the agency, but the [attractors] are effectively determined from the Big Bang."[19]

The second point to be made arises from this insight. It is that in speaking here in terms of teleology, I am not adopting the sort of quasi-vitalistic framework in which the components of the universe are seen as being drawn toward an intended final end by some external agent or force. "Attractors," in the sense in which Conway Morris uses the term, do not literally attract through some kind of force or influence that they exert. They are simply likely outcomes of laws of nature acting on the components of the universe, and may be understood, scientifically, in terms that make no reference to these outcomes themselves. We may, in a theological context, choose to speak of the reality of these attractors in terms of God's design of the entire universe, but if we do this, it is important to recognize that the tendencies we identify as part of the divine design are, at the level of scientific description, still absolutely intrinsic to the components of the universe and to the laws they obey. Theological interpretation of these tendencies in no way impinges on their scientific description in the way it would if a vitalistic understanding were adopted.

The relevance of this second point to the question of divine action becomes clear when we recall the character of the teleological tendency posited by the strand of the Byzantine tradition embodied in the work of Maximos the Confessor. For there, too, as we have seen, there is an understanding of the cosmos's teleological tendency that has precisely this nonvitalistic character.

Recognition of this parallelism between ancient and modern perspectives cannot, of course, lead in any simplistic way to the claim that the earlier model anticipates an important aspect of contemporary science. Clearly, at the level of details, this is far from true. At a more fundamental level, however, we can surely recognize that there is a broad consonance between the two kinds of understanding. By pointing to the way in which the "laws of nature" perceptible to the scientist have a teleological effect—both in the physical development of the cosmos and in the biological evolution of the species of our planet—scientific perspectives do suggest important parallels between what we now call the laws of nature and what Maximos the Confessor called the *logoi* of created things. At the very least, there seems to be a sense in which, when teleology at this low level is discussed, there need be no dissonance between scientific perspectives and the basic insights of the teleological-christological model that he articulated.

It can hardly escape our notice, however, that while the traditional Byzantine vision of the teleological aspect of the created order is somewhat obscure, it is clearly not limited to this low level of effect. Does this mean, then, that the parallelism between the modern concept of the laws of nature and the ancient one of the *logoi* of created things breaks down at this point?

Our answer to this question will depend largely on our judgment of the effects of complexity, which we have thus far considered in this chapter only in the context of a strong theistic naturalism. If we accept, in that context, that there is—as I have argued—no need to limit the universe's "fixed instructions" to scientifically explorable ones, then the same conclusion will apply here. This will mean that we see no reason to limit the teleological tendency of created things to the "inherent creativity" of the particular "laws of nature" that scientists can investigate. Rather, from the perspective of a teleological-christological model, it is quite possible to see the laws of nature that are perceptible to the scientist as representing no more than a "low-level" manifestation of what Maximos calls the characteristic *logoi* of created things. Over and above such manifestations, for this model, there may be, at least in principle, higher levels of manifestation that will—even though they are "lawlike"—inevitably be beyond what the scientific methodology is able to examine.[20]

To acknowledge that something may be possible in principle is not, of course, an adequate reason for accepting it as a persuasive model. We clearly need to ask ourselves, first, what these manifestations of a higher teleology are likely to be and, second, whether what we conjecture in this context has any correspondence to experience. Here, I would argue, the observable hierarchy of complexity in the cosmos is particularly relevant. For in the view of many, the scientific perspective points to the successive emergence in the cosmos of a number of levels of complexity—in particular, life and intelligent self-consciousness—that can be understood neither through a vitalistic approach nor through an ontological reductionism. Each of these emergent levels of complexity has, according to this view, an autonomous character that is explicable in terms of supervenience—often (but not always) expressed in terms of holistic organizing principles.[21]

Among those who have developed this view is Arthur Peacocke, whose particular philosophical arguments are comparable to those of many others. In one respect, however, his use of these arguments is distinctive, in that he suggests that the hierarchy of complexity that is perceptible through scientific observation should be understood theologically in a broader perspective. The religious aspect of human experience arises, he argues, from "the most integrated level or dimension that we know in the hierarchy of relations,"[22] so that theology as a discipline is related to the human sciences in much the same way as those sciences are to biology, and biology is to physics and chemistry.

The importance of this insight here is its implicit suggestion that, just as biology is rooted in physical phenomena, and specifically human qualities are rooted in biological ones, so we must see human religious experience as rooted in (but not reducible to) such aspects of the specifically human as the psychological.[23] When viewed in terms of the sort of teleological model I have outlined, this understanding suggests that human knowledge of God may be seen in terms of a psychological development that is a manifestation of the teleologically driven evolution of created things. Like life and like the specifically human qualities, revelatory experience of God may be seen as an aspect of the "natural" potential of the cosmos from the beginning. It, too, is an "attractor" in the sense in which that term is used in describing evolutionary convergence.

If we are to speak in this way, we must, however, be careful that the empirical, historical dimension of knowledge of God is taken fully into account. In particular, we must recognize that knowledge of God arises from revelatory experience, which is diverse in its manifestations and is not, when it occurs, always equally valid or valuable. If we are to speak of revelatory experience in terms of "teleologically driven evolution," we need some way of accounting for these factors.

Part of any such account will, inevitably, point toward scientific parallels. Just as in biology the notion of evolutionary convergence does not imply one single point of convergence but many, so a naturalistic understanding of revelatory experience should recognize the equivalent of a number of what Simon Conway Morris calls "stable nodes of functionality."[24] However, what these nodes will be in relation to revelatory experience is far from clear if we limit ourselves to biological insights. Here, I believe, the pansacramental approach to revelatory experience that I have advocated is important, since its way of positing a God-given "natural" tendency toward spiritual development provides an explanatory framework that, while using a biological analogy, transcends biology as such.

In this approach, as we have seen, the use of the term *pansacramental* does not imply that the potential of all created things can be actualized independently of context. Just as the specific sacraments of the church can be effected only in a particular ecclesial context, and the emergence of a new species requires a specific ecological niche, so also, I have argued, any particular revelation of God will take place only in what I call an appropriate psychocultural niche.

One aspect of such a niche is, as we have seen, a particular sort of psychological openness to God, since as Karl Rahner has emphasized, there is an essentially contemplative dimension to revelatory experience, from which arise the thoughts or visions that constitute the outer form of that experience.[25] (This does not, as we have noted, imply that revelatory experience necessarily takes place independently of associated empirical events, even apparently miraculous ones. It suggests, rather, that such events can be understood fully only in terms of the associated psychological states of those who experience them.)

This idea may be developed, I have argued, in terms of what I call a psychological-referential model of revelatory experience, which is related to important strands of thinking about revelation in the work of Rahner and of Yves Congar. Any authentic revelatory experience of God that occurs through this divinely given "natural" tendency always, I have suggested, takes a form appropriate to a particular cultural and psychological environment. Because of this, it can never be absolute. It may have a genuinely referential component, recognizable in principle through the "puzzle-solving" character of the theological language to which it gives rise.[26] However, it also inevitably has a culturally conditioned instrumental component, which makes the link between experience and referential doctrine complex. Moreover, I have noted, not only does this understanding include no a priori reason for believing that genuine revelatory experience can occur only among members of some particular religious grouping, but a number of factors suggest rather strongly a pluralistic understanding of the faiths of the world.

This pluralistic possibility—originally formulated in the context of a strong theistic naturalism—relates in an extremely interesting way to the neo-Byzantine model I am now suggesting. For in the early Byzantine model, from which the new model draws its inspiration, there was at least implicitly a comparable potential for pluralistic expansion. In particular, during the early patristic period, there was, as Philip Sherrard has noted, a positive attitude toward non-Christian religions among some Christian authors, based precisely on the kind of *Logos* understanding that we have noted. What is now needed, according to Sherrard, is a reappropriation of this approach.

Such a reappropriation is, I would suggest, made easier by the neo-Byzantine model I am advocating. This model may not actually entail the sort of pluralistic theology for which Sherrard makes his plea, which "affirms the positive attitude implicit in the writings of Justin Martyr, Clement of Alexandria, Origen, the Cappadocians, St. Maximos the Confessor and many others." When expanded as I have suggested, however, it certainly suggests an understanding in which, as Sherrard puts it, "the economy of the divine Logos" should not be "reduced to His manifestation in the figure of the historical Jesus." Rather, that economy will be seen in terms of a ubiquitous presence in which "the types of His reality" will be "equally authentic" whether found within the Christian world or outside of it.[27]

This pluralistic understanding is not, as I have indicated, an intrinsic component of the neo-Byzantine model of divine action that I have outlined. It is simply an auxiliary hypothesis[28] that is worthy of serious attention. Intrinsic to the model are, rather, a number of factors I have already outlined, with which, in summary, I now conclude this chapter.

My main point has been that a teleological-christological model allows us both to acknowledge the general insights about teleology that arise from the natural sciences and to appropriate these insights in such a way that we can avoid the conventional distinction between general and special providence. On the one hand, we can acknowledge that the teleological traits of the cosmos that are visible to the scientist —those to be seen in the physical development of the universe and in the biological evolution of the species of our planet—represent an important clarification of what we might call the "low-level teleology" inherent in a teleological-christological understanding. On the other hand, we can insist, from a theological perspective, that manifestations of the "higher-level teleology" will be inherent in that model. These latter manifestations, while lying beyond what the scientists' methodology can investigate, need not in any way be contradicted by a scientific understanding. They can, in principle, account for all that has previously been attributed to God's special providence.

Interpreted in this way, the teleological-christological model of divine presence and providence clearly manifests four advantages over competing models of divine action:

1. The model is based on an explicitly theological understanding, rather than on abstract philosophical questions about divine agency.
2. Questions about how God acts "on" the world—as if from outside—are rendered meaningless, since the model rejects the conceptual picture of what the cosmos can do "on its own" or when merely "sustained in being." This means, among other things, that the conventional distinction between general and special providence cannot be made, and all aspects of providence are comprehensible in terms of a single, simple model.
3. While the model is at one level "naturalistic," there need be no inherent limitation to the scope of divine providence, of the sort assumed by the deists and more recent advocates of a strong theistic naturalism. The question of what God has done or could do becomes not an abstract philosophical question, but a broader theological one, focused on a *Logos* Christology. Such a Christology, while it draws its inspiration from the Byzantine tradition, needs to echo that tradition only insofar as it must incorporate the sort of teleological understanding we have examined. In principle, an understanding of this sort might be developed in a number of ways, some of which, as we have noted, could reflect an essentially pluralistic position.

4. The model removes the tension between scientific understanding and belief in divine action in two distinct ways. It enables us to incorporate, within a theological perspective, specific aspects of scientific understanding that are sometimes held to challenge religious belief.[29] It also, in an important way, allows the intrinsic limitations of the scientific methodology to be seen much more clearly than hitherto.

16

Praying to the God beyond Time

At this point in my argument, I am tempted to stop. Certainly, if I were an advocate in a court of law, primarily motivated to convince a jury, then I might be well advised to do so. For in attempting to expand a strong theistic naturalism into a neo-Byzantine, teleological-christological model of God's action in the world, I have said all that is likely to be effective to persuade most readers that this model is preferable to its two obvious alternatives: that which adopts a simplistic theistic naturalism of a deistic kind, and that which adopts a two-mode model that makes a conventional distinction between general and special providence.

However, not only would stopping at this point be the equivalent of leaving out some of the evidence for my case, on the grounds that it risked making that case less convincing to the less subtle-minded juror, but it would also, and more importantly, leave only partially explored something that is essential to understanding the implications of my model for our general spirituality, particularly for the question of what we are doing when we pray for others or for ourselves. Therefore, although I may do so at my peril, I am about to outline a factor that, for me at least, is essential. The issue of intercessory prayer is important enough to make the risk worth taking.

The issue in question is the relationship of God to time, which has been at the heart of much of the recent discussion of divine action. For a significant factor in the way in which that discussion has progressed, especially within the dialogue of science and theology, has been the widespread abandonment of the traditional notion—consonant with important aspects of mystical experience[1]—that God's experience of time is fundamentally different from our own. For Thomas Aquinas in the thirteenth century, as for many in the previous centuries, God was to be seen as "wholly outside the order of time," so that "the whole course of time is subject to eternity in one simple glance."[2] Reacting against this traditional notion, however, many—on the grounds that a "personal" God must be able to "respond" to intercessory prayer—now insist that God must be seen as experiencing a past, a present, and a future in much the way that we ourselves do.

The prevalence of this assumption among those who have developed the new theology of nature is, in one way, very surprising, since their main claim to attention lies precisely in their attentiveness to scientific insights. One might have thought, therefore, that they would have paid more attention to one of the prime insights of modern physics: that any picture of a universe that develops within a uniformly unfolding temporal process is invalid. Indeed, for the scientifically aware observer, one of the most puzzling aspects of the tenacity with which a temporal view of God's experience is often held relates not to specifically theological considerations, but precisely to these scientific ones. From a scientific perspective, the "commonsense" view of the nature of time—in which there is a universal past, present, and future, which both God and God's creatures can experience—simply is no longer tenable.

This view was, admittedly, an intrinsic assumption of the Newtonian physics that held sway from the late seventeenth until the early twentieth century. In 1905, however, Albert Einstein published his work on what was later to be called the special theory of relativity, which set out an entirely new concept of the dynamics of bodies. Central to this new theoretical framework was the argument that the passage of time is not the same for all observers, but depends upon their velocities relative to one another. This has since then been corroborated experimentally in numerous ways, such as taking ionized atoms that can be used as atomic clocks, accelerating them to a high velocity, and noting the effect on their "clock" readings.

Moreover, Einstein's theory notes that not only does the time interval measured between two events depend on the different velocities of their observers, but under certain circumstances, even the order of the two events will be different for different observers. (This can happen when no signal traveling at the velocity of light or less could have provided a possible causal link between the events.) This means, among other things, that events that are in the past for some possible observers are in the future for others. There is no universal present.

While it is true that Einstein himself went on, a few years after the publication of this theory, to develop his general theory of relativity, which took these notions into new conceptual realms by incorporating also the effects of acceleration and gravity, the special theory's basic demolition of the commonsense notion of time was unaffected. The newer theory (as is usually the case in authentic scientific development[3]) simply allowed scientists to augment the correct predictions made by the old theory with new predictions applicable to situations where the old theory did not apply. Just as in the old theory, the notion of a universal present cannot be maintained.

From these relativistic theories arises a major insight: It is only because our human situation is one of relatively low velocities and weak gravitational interactions that our commonsense apprehension of space and time works for us at an everyday level. As the many experimental corroborations of relativity theory indicate, space and time are not in fact "given" and absolute, within which the created order unfolds. Rather, in relativistic theory, space and time become simply a part of the created order and are affected by other aspects of that created order.

In the context of a relativistic understanding, then, such concepts as past, present, and future—while not irrelevant when we think of such things as causality—become far from straightforward. Therefore, when we think of God's transcendence of the created order, we need to think not simply of God's transcendence of "things" as we usually understand them. And if, on the basis of scientific insight, we accept that what God transcends includes space and time themselves, then we seem to be already close to the classical view of divine eternity. At the very least, it seems impossible to defend the notion of God having a "present" in such a way that the past of all God's creatures is also "God's" past, so that God can "respond" to what has happened in that past. If defenders of a temporal picture of God are scientifically literate, then they must recognize that the events of our immediate "past" are still in the "future" for some possible observers, so the commonsense notion that God responds to our immediate past from a present that is identical to our own is valid only if we can insist that our frame of reference is more real to God than is that of these other observers. This insistence, it seems to me, is indefensible.

Some of those who have attempted to defend the notion of a temporal God have done so by implying that to speak of a nontemporal God is to deny the very reality of our experience of time. This they often see as linked in some way to the "block universe" interpretation of certain scientific insights,[4] which suggests the fundamental unreality of time.

Scientifically, as these people sometimes point out, there are good reasons for rejecting this sense of the meaninglessness of our subjective sense of time's flow from past to future. Suppose, for example, that a box is divided in half by a barrier, and the gas in one half of the box is pumped out. If the barrier is then removed, what happens to the ensemble of molecules in the other half is not reversible, even though what happens to each of the individual molecules is. When molecules start in just one half of the box, the result is—at least with very high probability—that the molecules will soon be more or less uniformly distributed throughout the box's volume. Conversely, if we start with a box full of gas molecules, then the probability is negligible that all the gas molecules will at some point in time come to lie in only half of the box. What the thermodynamicist calls entropy—broadly, the degree of disorder within any closed system—simply does not decrease as time goes on. The directionality of time indicated by this increasing entropy is a genuine aspect of the physics of the universe.

To acknowledge this is certainly to acknowledge that some kinds of interpretation of the "block universe" notion are not tenable. Contrary to the opinion of some commentators, however, this conclusion does not affect the classical notion of divine eternity in the least. For that notion is not, as is often suggested, intrinsically connected to the block universe concept. The classical view does not claim that time is unreal, but simply that our understanding of how God relates to it must take into account a number of different considerations, which are related, primarily, not to how time should be seen by us from within our space-time universe, but to how it is seen by God in God's transcendence of that universe. To the extent that the physics of time does affect our judgment of this issue, block universe notions—which are not even an intrinsic aspect of relativistic theory, although they are often linked with it—are of little consequence. What is important at this level is what is truly intrinsic to relativistic theory: the demolition of the concept of a universal past, present, and future shared by God and God's creatures.

Despite these considerations, however, there has been considerable resistance to what has been called the "classical view of divine eternity."[5] A number of complex factors have been instrumental in this, and many of these have been excellently summarized and analyzed in Antje Jackelen's study *Time and Eternity*.[6] One aspect of the situation that Jackelen does not focus on is, however, central for me. This is the common abandonment of the traditional apophaticism that is, as we have seen, characterized by Vladimir Lossky as transforming "the whole of theology into a contemplation of the mysteries of revelation."[7]

This apophatic attitude is, for example, central to our understanding of the (so-called) Athanasian Creed, with its proclamation of "The Father incomprehensible, the Son incomprehensible, and the Holy Ghost incomprehensible; the Father eternal, the Son eternal, and the Holy Ghost eternal," and with its further assertion that "they are not three eternals but one eternal, as also there are not three incomprehensibles nor three uncreated, but one uncreated and one incomprehensible."[8] It may be true that the relegation of this creed to the lumber room of liturgical history is in some ways fortunate, since few of us nowadays are happy with its declaration that the beliefs it outlines are such that, "except every one do keep [them] whole and undefiled, without doubt he shall perish everlastingly."[9] Nevertheless, this creed's imagery indicates something of immense importance. For speaking without frivolity about the Trinity as "not three incomprehensibles . . . but one incomprehensible" implicitly acknowledges the truth of Lossky's statement that "theology will never be abstract, working through concepts, but contemplative: raising the mind to those realities which pass all understanding. This is why the dogmas of the Church often present themselves to the human reason as antimonies, the more difficult to resolve the more sublime the mystery they express."[10]

The doctrine of the Trinity is, for Lossky, "pre-eminently an antimony"[11] If, he says, we speak of it in terms "of processions, of acts, or of inner determinations," then "these expressions—involving, as they do, the ideas of time, becoming and intention—only show to what extent our language, indeed our thought, is poor and deficient before the primordial mystery of revelation." Only a fully apophatic attitude, he insists, can allow us to "rid ourselves of concepts proper to human thought, transforming them into steps by which we may ascend to the contemplation of a reality which the created intelligence cannot contain."[12]

This is important in the context of our present exploration because many of our Christian forebears, with their acknowledgment of the incomprehensibility of God, would undoubtedly have been wary of attempting to reduce God's "personal" nature to what they themselves found comprehensible through their own experience of being persons. Such a reduction is, however, now very common, and appeal is made, by way of justification, to what is sometimes called the principle of the "analogy of being," according to which aspects of God can be clarified by appealing to aspects of human experience.

This principle is not, of course, entirely without its uses; in fact, I have already used it for a purpose that I believe is legitimate: that of emphasizing, in the context of models of divine providence, that what is possible for humans is not impossible for God. But those who appeal to the principle often go well beyond this kind of usage by suggesting, at least implicitly, that what is possible for God is in effect limited to what would be possible for humans if their existing knowledge and power were considerably increased. Thus, for example, it is often held that our experience of being human persons imposes strict limits on the ways in which God's "personal" involvement in the world can be effected.

In terms of the Trinitarian theology enshrined in the Athanasian and other creeds, however, it is very far from clear how our experience of being human persons can clarify our understanding of God's "personal" action. Just as prayer, for the early Christians, was always "to the Father, through the Son, and in the Holy Spirit," so all of God's gracious actions were "from the Father, through the Son, and by the Holy Spirit." God's action was not seen as "personal" in any facile sense of the term, but as tripersonal in an extremely subtle way. (One way of expressing this was, in fact, precisely that which I have advocated.) It is a sad fact, however, that Trinitarian considerations of this kind have been almost entirely ignored in recent debate about divine action, with the result that many have insisted on the centrality of a very simplistic concept of the "personal" nature of God.

This is brought out in what is often regarded as one of the most coherent arguments presented for a "temporal" God, which hinges on the assertion that "to deny the temporality of God is to deny that he is personal in any sense in which we understand personality."[13] The clue to what is going on in this argument lies, I suggest, in the precise form of wording used by its presenter, the philosopher J. R. Lucas. For the issue is not, in my view, whether he is correct in thinking that the classical view is incompatible with a God who is "personal in any sense in which we understand personality." (This is arguable, though by no means as evident as he seems to think.) The real issue is, rather, the extent to which human conceptual limitations should control our theological thinking. For Lucas, this control is implicitly accepted. The widespread inability to see how a nontemporal God can be compatible with God's being "personal in any sense in which we understand personality" is seen as definitive for our understanding of how God relates to temporal processes. For those for whom the apophatic tradition remains important, however, this inability need be no more than a sign that, in certain important respects, God is—as the Athanasian creed insists—"incomprehensible."

To say this is not to suggest that the concept of God's personhood is always inappropriate. As John Polkinghorne has often said, one of the things that he is attempting to do by stressing God's personal character is to avoid any concept of an "impersonal" God. (God's action must, he says, be seen as being more like that of a person than like that of, say, the force of gravity.) This, of course, we can accept. What apophaticism demands, however, is a very careful use of concepts such as personal and impersonal, and a recognition that both may be necessary to an apophatic approach that acknowledges the validity of antimony in theological language usage. God is, for the Christian, certainly not less than personal, in the everyday meaning of that term. In traditional Christian thinking, however, God is also far more. God is suprapersonal in a way that makes appeal to God's "personal" character potentially misleading. The antimony of speaking of God's being simultaneously personal and other than personal is essential to a valid theology and spirituality.

Thus, for example, when Polkinghorne makes his appeal to the "personal" nature of God in the context of his thinking about divine action, we need to look at his arguments with a critical eye. He has suggested, for example, that accounts with a greater stress on naturalistic perspectives than his own involve "an implicit deism . . . whose nakedness is only thinly covered by a garment of personalized metaphor."[14] This leads him to feel misgivings—echoed by others, including Philip Clayton[15]—even about the moderate degree of naturalism in the approach of someone like Arthur Peacocke.

The origin of this misgiving lies, it would seem, in an aspect of Peacocke's thinking that Polkinghorne has perceptively noted. For, despite recognizing how "unhelpful" the distinction between creation and providence often proves to be, Peacocke has, as we have seen, usually felt it necessary, when speaking of providence, to adopt a temporal, "response" model. This is necessary, in his view, in order to focus "upon *particular* events, or patterns of events, as expressive of the 'purposes' (e.g., of communication) of God, who is thereby conceived of as in some sense personal."[16] What Polkinghorne has noted, however, is that this aspect of Peacocke's thinking sits somewhat uneasily with many of its other aspects.[17] A good deal of Peacocke's approach, he suggests, is in fact more consonant with the "single act" approach to divine action, which stresses God's nontemporal nature, than with the way of thinking in which God is a temporal being who responds in a literal sense to events in the world.

Given this tension in Peacocke's thinking, we must ask why he has eschewed the nontemporal approach with which so much of his thinking seems to be in harmony. The answer seems to be simply that he has not found a way, within the nontemporal approach, of affirming God's "involvement" with the world, as expounded in the work of theologians who have influenced him, such as Jürgen Moltmann and W. H. Vanstone.[18] Therefore, like Polkinghorne and many others, he has adopted an approach to divine action that relies on our own experience of time in order to speak about divine providence. At least implicitly, it is assumed that general providence is based on decisions that God has made "in the past" as creator, while special providence results from God's "subsequent" decisions, which are "responses" to events in the world.

To see a nontemporal model of God as precluding God's "involvement" in the world, however, misreads its implications. Theologically, just as the classical idea of divine eternity can be expressed in terms of God's transcendence of created things, so God's "suprapersonal" involvement in temporal processes can be thought of in terms of God's immanence. Indeed, in terms of the kind of panentheistic model I have outlined, this immanence and involvement may be seen even more clearly than is the case in other frameworks to which these concepts are central. For if we see the divine *Logos* as present at the very heart of all created things, then the temporal experience of all created things—including their suffering—may be seen as nothing less than *God's own experience.*

This has important implications for the problem of evil, since for this kind of model, God may be seen not just as a sympathetic observer of the experiences of created beings, but as being quite literally at the very heart of those experiences. In this perspective, God does not just suffer *with* us in the way that those who posit a "suffering God" usually assume—that is, in a way analogous to that in which parents suffer in the presence of their pain-wracked child. Rather, for this model, God's immanence is truly radicalized: the suffering of any of God's creatures *is* God's suffering. God's pain is not the "personal" suffering of the sympathetic observer, but the "suprapersonal" pain of one who is closer to the suffering even than the consciousness of the person who suffers.

In God's transcendence of time, however, God must be seen as experiencing this suffering not only as do the creatures in and through whom God suffers, but in another way also. For in this model, the suffering of God's creatures is not apprehended by God only as a series of temporal sufferings, even though all that belongs to the temporal nature of those sufferings is truly God's. Rather, because God transcends time, all pain is, for this model, not only suffered by God within the temporal process, but also eternally present to God as transcendent God, since "the whole course of time is subject to eternity in one simple glance."

In this understanding, God does not "change" because of changes in the suffering of the world. In this sense God is, as traditional theology has insisted, "impassible." To affirm this is not, however, to imply that our sufferings are not part of God's own experience. Rather, just as traditional Christian theology affirms that the God who cannot change or suffer truly did, in the historical Jesus, suffer the pain and desolation of the cross, so here this traditional notion is universalized. Insofar as God is at the heart of our temporal experience, God does experience the physical and mental anguish that accompanies each new pain. It is, however, in terms of God's eternal experience of suffering—including that which is yet to come—that the temporal element of this suffering finds its meaning and ultimate justification.

In a similar way, while God does not "respond" to the world as a temporal being according to this scheme, this does not mean that God is not directly involved in all the events that we speak of as providential. It is true that the classical view of divine eternity has often been expressed, in the context of discussions about divine action, in impersonal terms. That the classical view need not necessarily be interpreted in this way is, however, clear from the kind of panentheistic model I have advocated. For, as we have just observed, in this kind of model, God may be seen as being quite literally at the very heart of all that happens, experiencing events from within. Similarly, for this model, the effects of God's laws of nature *are* God's action. God is closer to them than one who has merely manipulated those laws as intermediary tools.

An important aspect of what this means can be seen, I believe, from the fact that one of the ways in which I differ from most advocates of a nontemporal view of divine action is that I do not believe that "temporal" models of divine action are simply alternatives to the classical one that happens to be mistaken. Rather, I take seriously the charge, made by some of their opponents, that users of the language of primary and secondary causes either (like Willem Drees[19]) limit the scope of divine action in a deistic way or else (like the neo-Thomists[20]) regard the way in which God's will is effected by natural causes as ineluctably mysterious. John Polkinghorne is right, I believe, to criticize the way in which descriptions in terms of primary causes and those in terms of secondary causes can only "walk past each other at different levels of discourse."[21] As a result of the weight of this charge, I accept that a "nontemporal" model of divine action can be fully persuasive only if the complementarity of the two levels of description can be shown to be more than a simple assertion.

I would argue, however, that this can be done, insofar as the neo-Byzantine model I have outlined provides genuine and new insights into the intrinsic relationship between primary and secondary causes. Indeed, it is precisely in this sense that I can describe my view as a version of a strong theistic naturalism, since *from a temporal perspective*, only an extended naturalism provides, in my view, an adequate complement to a nontemporal scheme of primary and secondary causes. I do not claim that a strong theistic naturalism, on its own, is adequate as a view of divine action. Rather, I see it essentially as a complement to the sort of scheme that the classical view of divine eternity allows.

What this means in practice can be seen in part from a specific example: that of C. G. Jung's attempt, from a psychological perspective, to affirm the significance of the sort of experience in which "coincidences" strike us as so extraordinary that we take them to be meaningful, even when we have taken into account the wishful thinking and bad statistics that bedevil discussion of them.[22] This kind of experience of "synchronicity" suggests, for Jung, the reality of what he calls an "acausal connecting principle"—a genuine link between the events whose juxtaposition seems meaningful, which is not understandable in terms of normal, temporal causality.

Interestingly, this concept of an acausal connecting principle often seems rather nebulous, even to people who do not wish to deny meaningful coincidence as an aspect of experience. This sense of nebulousness arises because we instinctively view the issue of causality from a temporal perspective. The concept of synchronicity thus seems not explanatory but descriptive. Correlation between events that are not linked by any simple progression from cause to effect seems, in this temporal perspective, intrinsically beyond the realm of rational discourse, and Jung's "acausal connecting principle" comes close to being an oxymoron.

It is, however, precisely here that a two-aspect model is valuable. For what is simply beyond analysis in terms of temporal causation can, in the nontemporal perspective, be affirmed in a subtle way. It is the temporal perspective that limits our identification of causality to events or processes that can be seen to have subsequent "effects" through some machinelike mechanism. In the nontemporal model, this limitation is transcended in a way that has sometimes been expressed in terms of God's "foreknowledge."

This can be illustrated by reference to the important question of how we should understand the purpose of intercessory prayer. If a purely temporal perspective is adopted, then God is seen as knowing what we pray for only after we have prayed for it. Therefore, that prayer can be effective only if there is some mechanism by which God can "respond" to our prayer. (This, as we have noted, is one of the main arguments behind the notions of a temporal God and of the necessity for some sort of model for special divine action.) In terms of the classical view of divine eternity, however, this picture is simply rendered void. In this view, God does not have to "wait" for our prayer to occur, but simply "sees" it eternally. Therefore, God's action in relation to the prayer can be affirmed without it being seen as action that is a "response" in the temporal sense of the term.

Interpreting this nontemporal model is difficult, however, since our instinctive tendency is to try using terms drawn from our experience of the temporal dimension of empirical reality. To speak of "fore"-knowledge, for example, will immediately imply for us knowledge of some event "in advance" of its occurrence, whereas its intention is to indicate that God's knowledge of events is neither in advance nor in retrospect. It is simply knowledge from the perspective of eternity: *sub specie aeternitas.*

An important aspect of this problem may be seen in the vexed questions about human free will that have often arisen from the biblical concept of "election." Some have understandably interpreted this concept in terms of a simplistic notion of predestination, since, from a purely temporal perspective, it is difficult to see how our exercise of will can truly be free if God already "foresees" what we will actually do. (As Ian Barbour has said of its neo-Thomist version, the idea of foreknowledge, "despite the subtlety of its elaboration and the recognition of its mystery, seems to end by denying the reality of contingency and freedom."[23]) For our time-bound instincts, the concept of election can be understood only in terms of our having been "predestined" to a particular end and thus constrained in such a way that our sense of free will is illusory.

From the perspective of the classical view, however, this problem simply disappears. For this view—at least in its subtlest form—does not think of God as "fore"-seeing or of individuals being "pre"-destined at all. In the nontemporal model, God simply "sees" all events throughout the course of time, including what our end will be. We are not constrained by the fact that God knows, *sub specie aeternitas*, what our choices are. While the temporal model creates an undeniable tension between God's foreknowledge and our freedom of choice, the biblical references to the elect become intelligible in the classical model.

In a similar way, as we have just seen, the problem of what we are doing in our intercessory prayer disappears when the classical view of God's eternity is accepted. It is not that God "foresees" our prayer—although this concept can perhaps give us a hint of how God can be thought of as "responding" to what has not yet happened. In terms of the classical model, God neither responds nor foresees in any literal sense of these terms. God simply acts and sees through a single linked action and vision. The connecting principle that links intercessory prayer and the situation to which it relates is thus, for this model, an "acausal" one in the sense in which Jung uses the term; it cannot be related to cause and effect in terms of either "response" or "anticipation."

This has one immediate and, I believe, immensely important implication for our life of prayer. It means that, while we cannot ask God to change what we know has happened in the past, we can properly pray about situations in the past about which we are partially ignorant. As Russell Stannard has put it, the concept of a non-temporal God allows intercession in relation to "events that have already happened, but about which we do not as yet know the outcome," because it allows us to see how "an all-knowing God will be aware that we will be offering up such a prayer later, and he will take that into account in determining the outcome."[24]

This possibility of praying about past events is in fact a wonderfully liberating notion, since—as psychotherapists are well aware—much of our desire relates not to the present and future but to the past. Our yearning is not simply that things will, in the future, be as we want them to be, but also that what we know about the past may have been ameliorated by factors about which we have no knowledge. (For example, we often long for the healing of those whom we have hurt, but with whom we have lost the sort of direct contact that would allow us to ask forgiveness and in other ways contribute to that healing.) To be able to offer the past in prayer in this way—asking that something may already have occurred—is a quite extraordinary psychological release.

This aspect of our life of prayer is not, however, strictly relevant to the main theme of this study. What is relevant is, as I have indicated, that the classical view of divine eternity allows a strongly naturalistic model of divine providence to be seen in an entirely new light. Any such model necessarily makes its initial claim to validity in the context of a methodological framework in which we restrict ourselves to the concept of temporal causality. Its claim, when taken in this context, is simply that it is preferable to the alternative models that can be developed within that framework. When considered in relation to the classical view of divine eternity, however, it may also be understood as representing no more than the temporal manifestation of God's eternal action.

This complementarity is, in my view, vital to any proper understanding of God's action in the world. Therefore, if the classical notion of God's relationship to time has hardly been mentioned in the previous chapters of this book, this is not because I consider it unimportant. Although nothing I have said in those chapters actually requires the classical view of divine eternity for it to be argued as valid, no model that fails to take into account this classical view can, in my judgment, be adequate. The model I have advocated is, arguably, more acceptable than either the special-providence or the deistic model, even if a purely temporal framework is assumed. However, only when the classical view of divine eternity is accepted as the overarching framework within which it is understood do I think it may properly claim to be definitive.

Afterword

The contemporary theologians who have most strongly influenced all I have said in this book fall into two categories. Some of them, including Vladimir Lossky, Panayiotis Nellas, and Andrew Louth, are commentators on the Greek patristic and Byzantine tradition. Their work is valuable primarily because it makes accessible aspects of classical theological thought that are, despite being virtually unknown in the Christian West, greatly relevant to the questions that confront us in the twenty-first century. Others fall into the rather different category of those who have attempted a new creative theology, which takes into account not only traditional perspectives but also the insights of recent times.

The two main figures I have quoted, who fall into this latter category, may well—for those who know their work—seem to be strange bedfellows. The first of them, Arthur Peacocke, is a biochemist by training, and because of this, he has brought to theology a particular scientific expertise. This has led him to argue persuasively that scientific insights inevitably affect our theological understanding. The second, Philip Sherrard, is in this respect almost as different from Peacocke as it is possible to imagine, for he has cast himself in the role of the child who shouts out that the scientific emperor's new clothes are in fact not clothes at all.[1]

As will have been clear from all that I have written, I am, in this particular argument, on Peacocke's side. Sherrard, it seems to me, makes a number of fundamental mistakes in his comments on the scientific enterprise, partly through a failure to distinguish properly between the different ways in which scientists understand the nature of their insights, and partly through a failure to distinguish properly between the motivation of the technologist, who wants to manipulate the world, and that of the pure scientist, who wants to understand it in a way that can, at times, be essentially contemplative.

However, despite this immense blind spot and others,[2] Sherrard remains for me one of the most stimulating theological writers of his generation. From a position based on a deep reading of the Eastern patristic tradition, he has managed, as we have seen, to throw light on many diverse topics. One of these, which has been central to part of my argument, is the question of how the Christian should understand the other faiths of the world. Here Sherrard has stressed the way in which the doctrine of the incarnation, when properly understood, indicates that the *Logos* of God, manifested in the person of Jesus, is "hidden everywhere, and the types of His reality, whether in the forms of persons or teachings, will not be the same outside the Christian world as they are within it."[3]

This insight means, for Sherrard and also for me, that other religious faiths cannot be dismissed as inauthentic in any simplistic way, since a "deep reading" of them may well turn out to be "a reading of the Logos, the Christ." This does not necessarily mean that all faiths are equal and that our own is just one among many. It does mean, however, that we must see our own faith, like any other, as an outcome of the way in which God's *Logos* operates at all times and in all cultures. God has, as Sherrard puts it, "at different times and in different places . . . condescended to clothe the naked essence of [the principles underlying all reality] in exterior forms, doctrinal and ritual, in which they can be grasped by us and through which we can be led into a plenary awareness of their preformal reality."[4]

To the best of my knowledge, Sherrard never made any comments on Yves Congar's understanding of God's self-revelation, and he was perhaps unaware of it. Nevertheless, I have argued that much of his understanding can be validly expressed in terms of Congar's view that all revelation is of God, and that what we can say about God as a result of this revelation is only very partial, since only at the *eschaton* will we come to know God as fully as created beings are able to. Whether we interpret Sherrard's incarnationally focused views through the filter of Congar's eschatologically focused ones or vice versa, we get a view of revelation that corresponds, at least in one respect, very closely to my own.

As we have seen, however, another aspect of the nature of revelation is vital. This is the integration of our understanding of revelatory experience into our more general understanding of divine action. Here I have taken my bearings, initially at least, from Arthur Peacocke's understanding of naturalistic processes as representing the way in which God has, so to speak, "designed" the world so that it can "make itself" in the extraordinary and fruitful way in which it has. This notion, I have argued, may be extended so as to develop a strong theistic naturalism quite different to that of the deists.

As we have noted, this "strong" theistic naturalism is somewhat different from the "weak" version of it advocated by Peacocke himself, in which what he calls providential divine action is seen to require that God "respond" to events in the world. Throughout this book, I have argued that this perception of the need for a divine response is in fact illusory. A strongly naturalistic pansacramental approach is already implicit in much of Peacocke's approach, and this, when made explicit, manifests enough similarity to aspects of Eastern Christian thinking for it to be incorporated into what I have called an incarnational naturalism, expressible in terms of a neo-Byzantine model of God's presence and action in the world. In terms of this model, I have suggested, God's presence and action in the world may be seen simply as two sides of the same coin, and the usual Western distinction between general and special modes of divine action simply becomes unnecessary.

At the heart of this model's avoidance of a deistic view of the scope of divine providence is a view that takes seriously the Eastern patristic notion that our world reflects our "fallen" state. It does not matter, I have suggested, whether we see the fallen state of the universe as a consequence of the misuse of human free will or as an anticipation of it. In either scenario, the important thing is that the world as we know it is both transparent to God's purposes and, at the same time, partially opaque to them. When, with faith, we look at God through the window of the world in which we live, God can certainly be perceived. Nevertheless, in certain respects, this perception is only, in Paul's words, "through a glass darkly" (1 Cor 13:12), since—as the problem of natural evil indicates—the world is not yet entirely as God wills it to be.

Here again, some of the perspectives that Sherrard has stressed become important, for as I have argued, it is only in and through the restoration of the world's "original" form in Christ that this opaqueness can be removed, so as to make the world fully transparent once again to the presence of the *Logos* of God in each of its parts. I have thus suggested, in the wake of Sherrard, that when an "outward and visible sign" becomes, in a sacrament, the bearer of an "inward and spiritual grace," what is happening is not that God has operated "from the outside" through some "special" mode of action. Rather, through the human invocation of divine grace, the grime of a fallen cosmos has been removed to reveal, in some particular thing, its true, pristine reality.

Similarly, I have suggested, when a miracle occurs, this same restoration has occurred. At one level, we may speak of this occurrence in terms of fixed instructions or, in quasi-Augustinian nomenclature, in terms of a "higher" level of the laws of nature. However, such philosophical terms merely indicate what is happening when the cosmos is enabled to manifest its underlying reality. When this reality is revealed in all its splendor, there can be no "natural" evil, but only the fullness of the well-being that is God's ultimate intention for us. In terms of this concept, I have suggested, the traditional Western distinction between special and general providence is not meaningless. It represents, however, not a distinction between modes of God's action, as is usually thought, but a distinction between the cosmos's empirical ("unnatural") and eschatological ("natural") states. Even here and now, the latter can be evoked from within by our response to God in faith.

In his own day, Maximos the Confessor—on whose thought my neo-Byzantine model is partially based—set out what has sometimes been called his "cosmic vision." He was able to do this in the way that he did because he had at hand a philosophical framework that seemed adequate to the task. The new cosmic vision I have outlined cannot, however, be set out in a comparable way at the present time. The philosophical framework that served Maximos so well is no longer acceptable to most of us without major modification, and nothing that we have at hand can be used in a straightforward way for the task that lies ahead. Therefore, any present articulation of a theologically motivated cosmic vision must be a rather more piecemeal one, of the sort that this book has attempted.

Even so, a more philosophically coherent articulation may be possible. Maximos was the heir to centuries of Christian wrestling with the philosophical categories available in the late antique world, so as to mold them into a form adequate to express, for that age, God's self-revelation in Christ. At the beginning of those centuries of molding, the problems of an inadequate philosophical language were as great as they are in our own day, and this fact can surely encourage us. Just as the early patristic writers did in their age, so we too, in our own, must begin the task of constructing a doctrinal framework with the inadequate tools and insights we have on hand. If this task takes as long to complete as did that which culminated in the work of Maximos, we should not be surprised.

As I have indicated, however, doctrinal statements—though important to preclude spiritually dangerous opinions—are not at the center of the Christian faith. Their role is not to circumscribe the truth of Christ, but simply to safeguard the possibility of the direct encounter with God that occurs in and through him. Not only are doctrinal formulas always to be treated apophatically, but they must also, as Congar stresses, be seen as actualizing their significance primarily in the liturgical *experience* of the truths that they express. To put it another way, the doctrinal and liturgical "exterior forms" of the church exist only in order to lead us to the contemplation of the divine principles underlying all reality. These exterior forms do not circumscribe those principles, but serve, as Sherrard has insisted, only to allow us to "be led into a plenary awareness of their preformal reality."[5]

In this context, as we have seen, Sherrard has spoken of the need to reexamine the other faiths of the world through a theology that reflects the attitude "implicit in the writings of Justin Martyr, Clement of Alexandria, Origen, the Cappadocians, St. Maximos the Confessor and many others."[6] I have, in effect, extended this plea by suggesting that the whole of our understanding of the cosmos and of its redemption needs to be reexamined, not only in terms of what is implicit in this incarnational tradition of thinking, but also in terms of the natural sciences of our time.

Precisely what is likely to contribute to this new synthesis of scientific insight and theological wisdom will, inevitably, be a matter of controversial judgment. Just as in the formulation of the patristic synthesis of the early centuries, certain avenues of exploration that at first seem promising to some will turn out to be dead ends. We must not forget, however, that those who advocate these ultimately unfruitful avenues of exploration will be as important to the new task as were the "heretics" of old to the formation of classical Christian doctrine. By and large, it is only through error that the church as a whole is enabled to clarify its perspectives. Heretics may be failed explorers, but they are explorers nonetheless. Their very failures contribute to the map we need.

We must not forget, either, that heresy can be identified only with hindsight. We may rightly wish to see heresy as deviation from the faith "delivered once and for all to the saints," but novelty as such is not what constitutes such deviation. This is clear from the fact that the victors at the Council of Nicaea, though defending the soteriological insights of their forebears, were certainly introducing a novelty by insisting on the formula that Christ is one in essence (*homoousios*) with the Father. Their opponents were not just the heretical Arians, but also—perhaps even mainly—Christian conservatives resisting a novel vocabulary. For many decades during the fourth century, these "semi-Arians" (as they are sometimes called) resisted the Nicene affirmation, not because their faith was unorthodox, but simply because their focus was on the novelty of the verbal formula rather than on the questions the formula had so magnificently managed to answer.

In a similar way, as John Behr has noted, it is now easy for us to allow "our very familiarity with the reflections of the Fathers and the results of the dogmatic controversies and conciliar resolutions to blind us to the focal point of those reflections and debates: Jesus Christ, the crucified and exalted Lord." He notes, in particular, how his fellow Eastern Orthodox have often manifested "very little serious engagement with Scripture, or the pre-Nicene period" and have instead started with what they think they already know and then looked back to the Fathers "simply to find confirmation." This, he argues, represents an extremely faulty methodology, which carries a great risk of misconstruing what these Fathers were saying. As he rightly notes, "If the questions being debated are not understood, it will be difficult, if not impossible, to understand the answers."[7]

In much the same way, I would argue, the conclusions I have come to here cannot be judged simply by comparing them with the doctrinal assertions of former ages. In the face of the questions that confront us now, we must not respond with a formulaic version of what we think we know, but with a conviction that we can develop a creative and faithful response in spiritual continuity with the experience of Christians throughout the ages. If we see the historical teachings of the church from within and enter into its inner spirit, then we can, as Timothy Ware, Bishop Kallistos of Diokleia, has put it, "re-experience the meaning of Tradition in a manner that is exploratory, courageous and full of imaginative creativity."[8]

This, at any rate, is the spirit in which I offer this book to the reader, and the spirit in which I hope it has been read.

Notes

Preface

1. Christopher C. Knight, *Wrestling with the Divine: Religion, Science, and Revelation* (Minneapolis: Fortress Press, 2001).

2. These articles are (in order of publication):

"Hysteria and Myth: The Psychology of the Resurrection Appearances," *Modern Churchman* 31:2 (1989): 38–42.

"An Authentic Theological Revolution? Scientific Perspectives on the Development of Doctrine," *Journal of Religion* 74 (1994): 524–41.

"A New Deism? Science, Religion and Revelation," *Modern Believing* 36:4 (1995): 38–45.

"Psychology, Revelation and Interfaith Dialogue," *International Journal for Philosophy of Religion* 40 (1996): 147–57.

"Resurrection, Religion and 'Mere' Psychology," *International Journal for Philosophy of Religion* 39 (1996): 159–67.

"Natural Religion Revisited: New Perspectives on Revelation," in *The Interplay of Scientific and Theological World-Views*, ed. N. H. Gregersen, U. Gorman and C. Wassermann (Geneva: Labor et Fides, 1998): 197–208.

"The Resurrection as Religious Experience," *Modern Believing* 39:2 (1998): 16–23.

"Some Liturgical Implications of the Thought of David Jones," *New Blackfriars* 85 (2004): 444–53.

"Theistic Naturalism and the Word Made Flesh: Complementary Approaches to the Debate on Panentheism," in *In Whom We Live and Move and Have Our Being: Panentheistic Reflections on God's Presence in a Scientific World*, ed. Philip Clayton and Arthur Peacocke (Grand Rapids: Eerdmans, 2004): 48–61.

"Divine Action: A Neo-Byzantine Model," *International Journal for Philosophy of Religion* 58 (2005): 181–199.

"The God Who Acts: Traditional Perspectives on a Current Dilemma," *Sourozh: A Journal of Orthodox Life and Thought* 101 (2005): 2–15.

"Naturalism and Faith: Friends or Foes?" *Theology* 108 (2005): 254–63.

"Natural Law and the Problem of Contraception," *New Blackfriars* 87 (2006): 505–14.

"The Easter Experiences: A New Light on Some Old Questions", Theology 110 (2007) 83-91

"Emergence, Naturalism, and Panentheism: An Eastern Christian Perspective," in Arthur Peacocke, *All That Is: A Naturalistic Faith for the Twenty-First Century: A theological proposal with responses from leading thinkers in the religion-science dialogue,* ed. Philip Clayton, 81–92 (Minneapolis: Fortress Press, 2007).

"The Christian Tradition and the Faiths of the World: Some Aspects of the Thought of Philip Sherrard", Theology (forthcoming).

"The Fallen Cosmos: An Aspect of Eastern Christian Thought and its Relevance to the Dialogue Between Science and Theology" (submitted).

Chapter 1

1. See Christopher C. Knight, *Wrestling with the Divine: Religion, Science, and Revelation* (Minneapolis: Fortress Press, 2001), 50.

2. A. R. Peacocke, *Theology for a Scientific Age: Being and Becoming—Natural, Divine and Human,* enlarged ed. (London: SCM, 1993), 119.

3. Richard Dawkins, *The Blind Watchmaker* (London: Longman, 1986); and Richard Dawkins, *Climbing Mount Improbable* (Harmondsworth, England: Viking, 1996).

4. For a questionable defense of "intelligent design" in which valid comments about reductionism are made, see, for example, Neil Broom, *How Blind Is the Watchmaker? Nature's Design and the Limits of Naturalistic Science* (London: Inter-Varsity, 2001). On the question of reductionism, see Knight, *Wrestling with the Divine,* 36–37; see also Arthur Peacocke, *God and the New Biology* (London: Dent & Sons, 1986), 6ff.

5. See, for example, I. Lakatos, "Falsification and the Methodology of Scientific Research Programmes," in *Criticism and the Growth of Knowledge,* ed. I. Lakatos and A. Musgrave, 91ff (Cambridge: Cambridge University Press, 1970). I do not, it should be noted, advocate Lakatos's approach in all its details. Nevertheless, it provides a useful account of the role of scientific argument.

6. For a brief account of this, see Steve Connor, "Eminent Biologist Hits Back at the Creationists Who 'Hijacked' His Theory for Their Own Ends," *The Independent* (London), 9 April 2002, 3.

7. It should be noted that this phrase *once in being* does not necessarily imply that the big bang with which our universe began should be seen as due to supernatural input. For example, if this "beginning" represents, as some suggest, a fluctuation in an already existing quantum vacuum, then this quantum vacuum—including the laws it obeys—represents something that is in being "prior to" the big bang. The beginning of time in the big bang does not, therefore, necessarily coincide with ontological origin. What is important is that any scientific explanation of "the beginning" must always presuppose some underlying physical reality that in theological perspective is created by God.

8. Although new arguments have arisen in relation to it, the most comprehensive account of this principle is still that of John D. Barrow and Frank J. Tipler, *The Anthropic Cosmological Principle* (Oxford: Clarendon, 1986).

9. See Knight, *Wrestling with the Divine,* ch. 1.

10. John Polkinghorne, *Science and Creation: The Search for Understanding* (London: SPCK, 1988), 15.

11. Nicholas Lash, "Observation, Revelation and the Posterity of Noah," in *Physics, Philosophy and Theology: A Common Quest for Understanding*, ed. R. J. Russell, W. R. Stoeger, and G. V. Coyne (Vatican City: Vatican Observatory, 1988), 209.

12. John Polkinghorne, *Reason and Reality* (London: SPCK, 1991), 80.

13. See, for example, Stephen J. Gould, *Wonderful Life: The Burgess Shale and the Nature of History* (New York: Norton, 1989).

14. See Richard Dawkins, *The Ancestor's Tale: A Pilgrimage to the Dawn of Life* (London: Weidenfeld & Nicolson, 2004), which notes that the venomous sting, for example, has developed independently at least ten times, and true flapping flight at least four.

15. This probability can be expressed in terms of the well-known Drake equation. The actual figures to be assigned to some of the terms of this equation are, however, very much a matter of guesswork, and some scientists' assertions that intelligent life forms must exist elsewhere in the universe are, as a result, manifestations less of sober scientific judgment than of hyperbole. To give one example: the tidal effect of the moon has clearly been considerably important in the actual evolution of life on earth, yet the existence of a moon with such a large tidal effect seems to have been the outcome of a freak collision between the proto-earth and another large body, of a sort that could have happened to very few planets in the universe. This factor is, however, rarely taken into account by those who attempt to assign a plausible value to the term in the Drake equation that deals with the probability of land animals evolving, which, in the earth's case, was strongly affected by this tidal factor.

16. Simon Conway Morris, *Life's Solution: Inevitable Humans in a Lonely Universe* (Cambridge, Cambridge University Press, 2003). I would argue, for example, that it is not unthinkable that intelligent beings might have arisen with the kind of echolocation mechanism that has arisen on Earth four times: in bats, toothed whales, oil birds, and cave swiftlets.

17. Paul Davies, *The Cosmic Blueprint* (London: Unwin Hyam, 1987), 143.

18. Philip Clayton, *Mind and Emergence: From Quantum to Consciousness* (Oxford: Oxford University Press, 2004), provides the best general survey of this issue. See also Dennis Bielfeldt, "The Peril and Promise of Supervenience for the Science-Theology Discussion," in *The Human Person in Science and Theology*, ed. Niels Henrik Gregersen, Willem B. Drees, and Ulf Gorman, 117ff (Edinburgh: T&T Clark, 2000).

Chapter 2

1. This phrase is associated with the sort of "postfoundationalist" understanding of scientific rationality that is outlined in Christopher C. Knight, *Wrestling with the Divine: Religion, Science, and Revelation* (Minneapolis: Fortress Press, 2001), ch. 5. In a theological context, this outlook has been explored (and perhaps slightly overstated) by people such as J. Wentzel van Huyssteen; see, for example, his essay "Postfoundationalism in Theology and Science," in *Rethinking Theology and Science: Six Models for the Current Dialogue*, ed. N. H. Gregersen and J. W. van Huyssteen 13ff (Grand Rapids: Eerdmans, 1998).

2. T. S. Kuhn, *The Structure of Scientific Revolutions* (Chicago: University of Chicago Press, 1962). See Knight, *Wrestling with the Divine*, 50ff.

3. See Knight, *Wrestling with the* Divine, ch. 5.

4. Michael Polanyi, *Personal Knowledge* (London: Routledge & Kegan Paul, 1958). It is interesting to note that even before Kuhn's analysis, Polanyi seems to have anticipated some of the responses to it.

5. Richard Dawkins, "The Great Convergence," reprinted in his *A Devil's Chaplain: Selected Essays by Richard Dawkins*, ed. Latha Menon (London: Weidenfeld & Nicolson, 2003), 173.

6. Ibid.

7. Ibid., 174.

8. Ibid., 173.

9. Keith Ward, *A Vision to Pursue: Beyond the Crisis in Christianity* (London: SCM, 1991), vii.

10. See Knight, *Wrestling with the Divine*, ch. 7.

11. Metropolitan Anthony of Sourozh, *Encounter* (London: Darton, Longman & Todd, 2005), 31.

12. See, for example, Imre Lakatos, "Falsification and the Methodology of Scientific Research Programmes," in *Criticism and the Growth of Knowledge*, ed. I. Lakatos and A. Musgrave, 91ff (Cambridge: Cambridge University Press, 1960). For a theological application of this approach, see Nancey Murphy, *Theology in an Age of Scientific Reasoning* (Ithaca: Cornell University Press, 1990).

13. For an account that stresses the dangers of taking this tradition of exegesis too far in the direction of ignoring historical perspectives, see R. P. C. Hanson, "Biblical Exegesis in the Early Church," in *The Cambridge History of the Bible*, vol. 1, *From the Beginnings to Jerome*, ed. P. R. Ackroyd and C. F. Evans, 412ff (Cambridge: Cambridge University Press, 1970).

14. See, for example, the quotation given in Olivier Clement, *The Roots of Christian Mysticism: Text and Commentary* (London: New City, 1993), 30–31.

15. Vladimir Lossky, *The Mystical Theology of the Eastern Church* (Cambridge: James Clarke, 1957), 33, referring to Gregory of Nyssa, *De Vita Moysis* (PG 44:377).

16. For example, in Gregory of Nyssa's *On the Making of Man*, he speaks in the fourth century about the way in which, through rest and motion,

> the Divine power and skill was implanted in the growth of things, guiding all things with the reins of a double operation. . . . These, moreover, were framed before other things, according to the Divine wisdom, to be as it were a beginning of the whole machine (I.1). . . . Of the corporeal [part of creation], part is entirely devoid of life, and part shares in vital energy. Of a living body, again, part has sense conjoined with life, and part is without sense: lastly, that which has sense is again divided into rational and irrational. For this reason the lawgiver [the author of Genesis, supposed to be Moses] says that after inanimate matter (as a sort of foundation for the form of animate things), this vegetative life was made, and had earlier existence in the growth of plants: then he proceeds to introduce the genesis of those creatures which are regulated by sense: and since, following the same order, of those things which have obtained life in the flesh, those which have sense can exist by themselves even apart from the intellectual nature, while the rational principle could not be embodied save as blended with the sensitive,—for this reason man was made after the animals, as nature advanced in an orderly course to perfection. (VIII.5)

17. Panayiotis Nellas, *Deification in Christ: Orthodox Perspectives on the Nature of the Human Person*, trans. Norman Russell (Crestwood, N.Y.: St. Vladimir's Seminary Press, 1997), 33.

18. Metropolitan Anthony of Sourozh, *God and Man*, rev. ed. (London: Darton Longman & Todd, 2004), 51–52.

19. See Knight, *Wrestling with the Divine*, ch. 5, for a description and critique of this Popperian approach, which in certain respect stands in marked contrast to that of Kuhn discussed earlier in this chapter.

20. Metropolitan Anthony, God and Man, 56–57.

Chapter 3

1. A. R. Peacocke, *Theology for a Scientific Age: Being and Becoming—Natural, Human and Divine*, enlarged ed. (London: SCM, 1993), 65.

2. Ibid., 119.

3. The popular picture that Christian reaction to Darwinism is based on a literalist reading of the scriptures is in fact a gross oversimplification. For example, many Christians had already accepted geological evidence for the great age of the earth.

4. A. L. Moore, *Science and Faith* (London: Kegan Paul, Trench & Co., 1889), 73.

5. W. G. Pollard, *Chance and Providence* (London, Faber & Faber, 1958).

6. See, for example, John Polkinghorne, *Faith, Science and Understanding* (London: SPCK, 2000), 120–21.

7. John Polkinghorne, *Scientists as Theologians: A Comparison of the Writings of Ian Barbour, Arthur Peacocke and John Polkinghorne* (London: SPCK, 1996), 41.

8. Ibid., 40.

9. A. R. Peacocke, "God's Interaction with the World: The Implications of Deterministic 'Chaos' and of Interconnected and Interdependent Complexity," in *Chaos and Complexity: Scientific Perspectives on Divine Action*, ed. R. J. Russell, N. Murphy, and A. R. Peacocke, 263–287 (Vatican City: Vatican Observatory, 1995), 283.

10. This means that the "whole" that is involved in any divine response is not the whole universe but a very small part of it. For God to "respond" to some event within time *t*, only events occurring less than a distance *ct* away (where *c* is the velocity of light) can be effective. Thus, for example, the "whole" that is involved if God is to respond through "whole-part constraint" in less than three years does not include even the nearest star (other than the sun), since this is more than three light-years away.

11. For some interesting comments on this from a different perspective, see T. F. Tracy, "Narrative Theology and the Acts of God," in *Divine Action: Studies Inspired by the Philosophical Theology of Austin Farrer*, ed. B. Hebblethwaite and E. Henderson, 173ff (Edinburgh: T&T Clark, 1990).

12. Polkinghorne, *Scientists as Theologians*, 41.

13. Arthur Peacocke, *All That Is: A Naturalistic Faith for the Twenty-First Century* (Minneapolis: Fortress Press, 2007), 20.

14. Nicholas Saunders, *Divine Action and Modern Science* (Cambridge: Cambridge University Press, 2002), 215.

Chapter 4

1. Christopher Bryant, *Jung and the Christian Way* (London: Darton Longman & Todd, 1983), 38.

2. C. G. Jung, *Flying Saucers: A Modern Myth of Things Seen in the Sky* (London: Routledge & Kegan Paul, 1959), 145.

3. David Bohm, *Wholeness and the Implicate Order* (London: Routledge & Kegan Paul, 1980), 197.

4. John Polkinghorne, *One World: The Interaction of Science and Theology* (London: SPCK, 1986), 74ff.

5. For comments on the biblical basis of this reappraisal, see Philip Clayton, *God and Contemporary Science* (Edinburgh: Edinburgh University Press, 1997). For interesting comments on the later use of the phrase "creation *ex nihilo*," see Philip Sherrard, *Christianity: Lineaments of a Sacred Tradition* (Edinburgh: T&T Clarke, 1996), ch. 10.

6. See, for example, Philip Clayton and Arthur Peacocke, eds., *In Whom We Live and Move and Have Our Being: Panentheistic Reflections on God's Presence in a Scientific World* (Grand Rapids: Eerdmans, 2004).

7. Willem Drees, "Thick Naturalism: Comments on Zygon 2000," *Zygon: Journal of Religion and Science* 35 (2000), 851.

8. Andrew Louth, "The Cosmic Vision of St. Maximos the Confessor," in *In Whom We Live*, ed. Clayton and Peacocke, 188.

9. Stephen W. Need, "Rereading the Prologue: Incarnation and Creation in John 1.1–18," *Theology* 106 (2003), 403.

10. Arthur Peacocke, *God and the New Biology* (London: Dent & Sons, 1986), 124.

11. Ibid.

12. See Christopher C. Knight, *Wrestling with the Divine: Religion, Science, and Revelation* (Minneapolis: Fortress Press, 2001), especially ch. 2.

Chapter 5

1. An interesting brief account is given, for example, in Bas C. van Fraasen, *The Scientific Image* (Oxford: Clarendon Press, 1980), ch. 5.

2. Augustine, *Of the Advantages of Believing* 34; cf. *City of God* 21:6–8.

3. See, e.g., Raymond E. Brown, *The Gospel According to John* (New York: Anchor, 1966, 1970), 530.

4. Ian T. Ramsey, *Religious Language* (London: SCM, 1957), 147.

5. For an expression of this view, see, for example, Raymond E. Brown, *The Virginal Conception and Bodily Resurrection of Jesus* (New York: Paulist Press, 1973).

6. Ibid., 132.

7. See, for example, Christopher C. Knight, "Jesus the Bastard and Mary the Mother of God," *Theology* 106 (2003), 411ff.

8. One argument of this sort stresses that the strands of Christianity that put little emphasis on Mary, the mother of Jesus, manifest a one-sided masculinity in their psychological effects. C. G. Jung, for example, hailed the Roman proclamation of the doctrine of her immaculate conception as a psychologically sound move in this respect.

9. This is partially expressed in Knight, "Jesus the Bastard."

10. Robert John Russell, "Bodily Resurrection, Eschatology, and Scientific Cosmology," in *Resurrection: Theological and Scientific Assessments*, ed. Ted Peters, Robert John Russell, and Michael Welker, 3ff (Grand Rapids: Eerdmans, 2002).

Chapter 6

1. One notable example of this is the strand of historical understanding that some scholars derive from the Q source with which Matthew and Luke embellished what they had derived from Mark (who may not himself, in the views of some, have known a specific

resurrection appearance tradition). For a popular account of this strand of thinking, see B. L. Mack, *The Lost Gospel* (Longmead, England: Element, 1993).

2. Norman Perrin, *The Resurrection Narratives: A New Approach* (London: SCM, 1977).

3. Christopher Knight, "Hysteria and Myth: The Psychology of the Resurrection Appearances," *Modern Churchman* 31:2 (1989), 38ff.

4. G. E. Selwyn, "The Resurrection," in *Essays Catholic and Critical—by Members of the Anglican Communion*, ed. G. E. Selwyn (London: SPCK, 1926), 296.

5. Hans Urs von Balthasar, *The Glory of the Lord: A Theological Aesthetics*, vol. 1, *Seeing the Form* (Edinburgh: T&T Clark, 1982), 415.

6. Karl Rahner, "Visions and Prophecies," in *Studies in Modern Theology* (Heidelberg: Herder; London: Burns & Oates, 1965), 97.

7. Ibid., 114.

8. Ibid., 118–19.

9. Ibid., 122.

10. Ibid., 124.

11. Ibid., 126.

12. Ibid., 127–28.

13. Christopher F. Schiavone, *Rationality and Revelation in Rahner* (New York: Peter Lang, 1994), 229.

14. L. O'Donovan, "Karl Rahner: Foundations of Christian Faith," *Religious Studies Review* 5 (1979), 198.

15. Schiavone, *Rationality and Revelation*, 238.

16. Rahner, "Visions and Prophecies," 138–39.

17. The so-called anthropomorphite controversy of 399 CE makes it clear that this belief was still alive at the end of the fourth century, although by then the majority of sophisticated believers thought it simplistic. The notion of God's form was not, however, a mere product of ignorance, but seems to have been related to earlier Christian and Jewish notions about God's glory. See, for example, Alexander Golitzin, "The Vision of God and the Form of Glory: More Reflections on the Anthropomorphite Controversy of AD 399," in *Abba: The Tradition of Orthodoxy in the West—Festschrift for Bishop Kallistos (Ware) of Diokleia*, ed. John Behr, Andrew Louth, and Dimitri Conomos, 273ff (Crestwood, N.Y.: St. Vladimir's Seminary Press, 2003).

18. See, for example, J. M. Robinson, "Ascension," in *The Interpreter's Dictionary of the Bible* (Nashville: Abingdon, 1962).

19. Ibid.

20. See, for example, the comments about the hope of resurrection in the Judaism of this period in D. S. Russell, *The Jews from Alexander to Herod* (Oxford: Clarendon, 1967), 148.

Chapter 7

1. See Christopher C. Knight, *Wrestling with the Divine: Religion, Science, and Revelation* (Minneapolis: Fortress Press, 2001), ch. 6.

2. Thus, for example, Jung sees the widespread "sightings" of UFOs in the twentieth century as an archetypal phenomenon whose specific form is related to a particular, technologically oriented culture. See C. G. Jung, *Flying Saucers: A Modern Myth of Things Seen in the Sky* (London: Routledge & Kegan Paul, 1959).

3. C. G. Jung, *Collected Works*, vol. 11 (London: Routledge & Kegan Paul, 1953–79), para. 10.

4. My own view is that Jung uses very interesting empirical foundations to develop a questionable theoretical superstructure.

5. For my own analysis of a few of these, see Knight, *Wrestling with the Divine*, ch. 4.

6. For an account of the "puzzle-solving" nature of the religious language that emerges from revelatory experience, see ibid., chs. 7 and 8, which provides one avenue of approach to this issue.

7. The way in which we approach this question will inevitably be influenced by our wider understanding of the resurrection of all at the end of the age. If, following the tradition of people like Tertullian, we emphasize the resurrection of the flesh of this mortal life, we are likely to stress the straightforward continuity between Christ's earthly and resurrection body. If we instead follow the tradition of writers like Gregory of Nyssa and Maximos the Confessor, who (following Paul) clearly distinguish the properties of the earthly and the resurrection bodies, we may take a somewhat different line. Certainly, these latter writers were influenced by an Origenist strand of thinking, which emphasized a "spiritual" rather than a "bodily" resurrection.

8. For an account of how this puzzle-solving aspect of religious language relates to the comparable puzzle-solving aspect of scientific language, see Knight, *Wrestling with the Divine*, chs. 7 and 8.

9. See ibid., ch. 5.

10. Quoted by W. Henn, "The Hierarchy of Truths According to Yves Congar, O.P.," *Analecta Gregoriana* 246 (1987), 115.

11. Ibid., 109–10.

Chapter 8

1. Philip Sherrard, *The Rape of Man and Nature* (Ipswich, England: Golgonooza, 1993), 51–52.

2. Philip Sherrard, *Christianity: Lineaments of a Sacred Tradition* (Edinburgh: T&T Clark, 1998), 61.

3. Ibid., 61–62.

4. Conversion to an existing religious faith (the "revelation" to an individual of its authenticity) is extremely common in comparison with the occurrence of foundational revelatory experiences. This argues that even if psychological factors are as important in the former as in the latter, they need not be as "finely tuned." For example, although the revelation in the person of Jesus Christ could only have happened within the psychocultural niche provided by early-first-century Judaism, that revelation clearly had features (which in a Jungian framework would be described as archetypal) that enabled its "truth" to be "revealed," in individual conversion experiences, not only to people within that culture, but also to many who inhabited different niches.

5. See, for example, Paul F. Knitter, *No Other Name?* (London: SCM, 1985), 145ff.

6. George Lindbeck, *The Nature of Doctrine: Religion and Theology in a Postliberal Age* (Philadelphia: Westminster, 1984).

7. See, for example, the comments in Derek Stanesby, *Science, Religion and Reason* (London: Routledge, 1985), 178–79.

8. See Christopher C. Knight, *Wrestling with the Divine: Religion, Science, and Revelation* (Minneapolis: Fortress Press, 2001), chs. 7 and 8.

9. Raimundo Panikkar, *The Unknown Christ of Hinduism*, rev. and enlarged ed. (London: Darton, Longman & Todd, 1981).

10. Keith Ward, *A Vision to Pursue: Beyond the Crisis in Christianity* (London: SCM, 1991), 175.

11. Sherrard, *Christianity*, 61–62.

12. The next two chapters, it should be stressed, do not represent an attempt to deal with this issue exhaustively, but simply indicate some of the issues involved in understanding religious language, some of which I have discussed in *Wrestling with the Divine*, chs. 6–9, and others of which I hope to explore in a future book-length study.

Chapter 9

1. Obituary of Philip Sherrard, *Times* (London), 6 June 1995.

2. Philip Sherrard, *Christianity: Lineaments of a Sacred Tradition* (Edinburgh: T&T Clarke, 1998), ix.

3. Ibid., xviii

4. Ibid., xviiif.

5. Sherrard, *Christianity*, x.

6. Obituary of Philip Sherrard.

7. Sherrard, *Christianity*, 53.

8. Ibid., 51.

9. Ibid., 56.

10. Ibid., 57.

11. Irenaeus, *Adversus haereses* III.12.13.

12. Origen, *Commentary on John* I.7.

13. Sherrard, *Christianity*, 58.

14. Ibid.

15. Ibid., 59–60.

16. Ibid., 61.

17. Ibid., 62.

18. Ibid., 63.

19. Ibid.

20. Ibid., 70.

21. Ibid., 71.

22. Ibid., 72–73.

23. Ibid., 73.

24. Ibid., 13–14.

25. Ibid., 73.

26. Ibid., 75.

27. Ibid., 61 (my italics).

28. Ibid., 62.

29. Ibid.

30. Keith Ward, *Religion and Revelation: A Theology of Revelation in the World's Religions* (Oxford: Clarendon, 1994).

31. John Behr, *The Mystery of Christ: Life in Death* (Crestwood, N.Y.: St. Vladimir's Seminary Press, 2006), 178.

32. Ibid., 174.

33. Ibid., 179.

34. Ibid., 17.

35. Brian Davies, *An Introduction to the Philosophy of Religion*, 2nd ed. (Oxford: Oxford University Press, 1993), 141.

36. Behr, *The Mystery of Christ*, 179.

Chapter 10

1. Philip Sherrard, *Christianity: Lineaments of a Sacred Tradition* (Edinburgh: T&T Clark, 1998), 63.

2. Andrew Louth, *St. John Damascene: Tradition and Originality in Byzantine Theology* (Oxford: Oxford University Press, 2002), 213.

3. Ibid., 216–17.

4. Ibid.

5. Ibid., 216, 218.

6. David Jones to René Hague, 9–15 July 1973, in *Dai Greatcoat: A Self-Portrait of David Jones in His Letters*, ed. René Hague (London: Faber & Faber, 1980), 249.

7. Harman Grisewood, ed., *Epoch and Artist: Selected Writings by David Jones* (London: Faber & Faber, 1959).

8. Maurice de la Taille, *The Mystery of Faith and Human Opinion Contrasted and Defined*, trans. J. B. Schimpf (London: Sheed & Ward, 1930), 212.

9. Jones, "Art and Sacrament," in *Epoch and Artist*, 163, footnote.

10. Ibid., 163.

11. Ibid., 165.

12. Ibid., 171.

13. Ibid.171–72

14. Ibid., 173.

15. Ibid., 174.

16. Ibid. 175–76.

17. Ibid., 176.

18. Ibid., 177.

19. A. M. Allchin, "A Discovery of David Jones," in *The World Is a Wedding: Explorations in Christian Spirituality* (London: Darton, Longman & Todd, 1978), 162.

20. See especially the preface to David Jones, *The Anathemata* (London: Faber, 1952), where he notes, "'Tsar' will mean one thing and 'Caesar' another to the end of time" (p. 13), and his letter to *The Tablet* published on 26 April 1958, and reprinted in *Epoch and Artist*, 260–61, in which he discusses the hymn *Vexilla Regis*.

21. *Dai Greatcoat*, 249.

22. Jones to Harman Grisewood, 6 July 1964, in *Dai Greatcoat*, 207.

23. Jones to René Hague, 8–16 June 1966, in *Dai Greatcoat*, 224.

24. Ibid.

25. J. A. Jungmann, *The Place of Christ in Liturgical Prayer*, trans. A. Peeler (London: Geoffrey Chapman, 1965).

26. Thomas F. Torrance, *Theology in Reconciliation: Essays towards Evangelical Unity in East and West* (London: Geoffrey Chapman, 1975), ch. 4.

27. Arguably, the classical liturgical action is not entirely obscured with the new orientation when, either architecturally or in the bodily actions of the celebrant (or preferably both), there is a strong *vertical* dimension to the symbolism of the eucharistic action. In practice, however, the horizontal priest-people (Christ-church) interaction is too often emphasized in orientational symbolism, rather than any offering to the Father.

28. *The Constitution on the Sacred Liturgy*, art. 59—see Austin Flannery, ed., Vatican II: the Liturgy Constitution (Dublin: Sceptre, 1964), 100.

29. Mark R. Francis, "Sacramental Theology," in *The Blackwell Encyclopedia of Modern Christian Thought*, ed. Alister E. McGrath (Oxford: Blackwell, 1993), 586.

30. Ibid.

31. The best brief introduction to this sociological perspective is perhaps P. L. Berger, B. Berger, and H. Kellner, *The Homeless Mind: Modernization and Consciousness* (Harmondsworth, England: Penguin, 1974).

32. George Lindbeck, *The Nature of Doctrine: Religion and Theology in a Postliberal Age* (Philadelphia: Westminster, 1984). For many, the main problem with Lindbeck's analysis is its essentially instrumentalist understanding of religious language. However, for an argument that this is not a necessary inference from his prime insights, see Christopher C. Knight, *Wrestling with the Divine: Religion, Science, and Revelation* (Minneapolis: Fortress Press, 2001), ch. 11.

33. Jones to René Hague, 8–16 June 1966, in *Dai Greatcoat*, 222.

34. Jonathan Miles and Derek Shiel, *David Jones: The Maker Unmade* (Bridgend, Wales: Seren, 1995), 241.

35. Jones to René Hague, 8–16 June 1966, in *Dai Greatcoat*, 224.

36. Miles and Shiel, *David Jones*, 7.

37. Ibid.

38. Ibid., 293.

39. The best introduction to this issue possibly remains F. W. Dillistone, ed., *Myth and Symbol* (London: SPCK, 1966). See especially the essay in it by Ian T. Ramsey, "Talking About God: Models Ancient and Modern," 76–97.

40. Louth, *St. John Damascene*, 217.

41. G. E. H. Palmer, Philip Sherrard, and Kallistos Ware, eds. and trans., *The Philokalia: The Complete Text Compiled by St. Nikodemos of the Holy Mountain and St. Makarios of Corinth*, 5 vols. (London: Faber & Faber, 1979–95, one volume in press), definition of "fantasy" in the glossary of each volume.

Chapter 11

1. See for example Charles E. Curran, "Natural Law and Contemporary Moral Theology," in *Contraception: Authority and Dissent,* ed. Charles E. Curran, 151–175, (London: Burnes and Oates, Herder and Herder, 1969).

2. Philip Sherrard, *Christianity and Eros: Essays on the Theme of Sexual Love* (London: SPCK, 1976), 40.

3. Vladimir Lossky, *The Mystical Theology of the Eastern Church* (Cambridge: James Clarke, 1957), 101.

4. Sherrard, *Christianity and Eros*, 8.

5. Ibid., 13–14.

6. Ibid.

7. Ibid., 14.

8. Ibid., 18.

9. Ibid., 25.

10. Ibid., 25–26.

11. Ibid., 26.

12. Ibid., 5ff.

13. Ibid., 76–77.

14. Ibid., 26 (italics mine).

15. See in particular Philip Sherrard, *The Rape of Man and Nature* (Ipswich, England: Golgonooza, 1993).

16. See, for example, the comments in Mary Midgley, *Beast and Man: The Roots of Human Nature*, rev. ed. (London: Routledge, 1995), 39.

17. Irenaus Eibl-Eibesfeldt, *Love and Hate*, trans. Geoffrey Strachan (London: Methuen, 1971), 143.

18. Eibl-Eibesfeldt (ibid.) notes in this respect how psychologists, while right in suggesting links between sexuality and other aspects of life, are often wrong in their explication of it. Thus, for example, after noting (n. 17) the link between sexual and parenting behavioral patterns, he comments that Freud, "in a strikingly topsy-turvy interpretation, once observed that a mother would certainly be shocked if she realized how she was lavishing sexual behaviour patterns on her child. In this case Freud got things back to front. A mother looks after her children with the actions of parental care; these she also uses to woo her husband" (143).

Chapter 12

1. See, for example, J. L. Mackie, "Evil and Omnipotence," in *The Philosophy of Religion*, ed. B. Mitchell (Oxford: Oxford University Press, 1971), 92–93.

2. F. R. Tennant, *Philosophical Theology* (Cambridge: Cambridge University Press, 1930).

3. For an example of this approach, see David Bentley Hart, *The Doors of the Sea: Where Was God in the Tsunami?* (Grand Rapids: Eerdmans, 2005).

4. John Hick, *Evil and the God of Love* (London: Collins, 1966), ch. 12.

5. For one such, which will not be commented on further here, see Archimandrite Vasileos of Stavronikita, *Hymn of Entry: Liturgy and Life in the Orthodox Church* (Crestwood, N.Y.: St. Vladimir's Seminary Press, 1984), 113, based in part on Pseudo-Dionysius, *On the Divine Names* 4:20 (PG 3:717C).

6. Panayiotis Nellas, *Deification in Christ: Orthodox Perspectives on the Nature of the Human Person*, trans. Norman Russell (Crestwood, N.Y.: St. Vladimir's Seminary Press, 1997), 33.

7. Ibid., 33–34.

8. Ibid., 34.

9. Ibid., 35, quoting Nicolas Kavasilas, *The Life in Christ* (PG 150:681AB).

10. Ibid.

11. Ibid., 37.

12. Ibid., 39.

13. Ibid., 41.

14. Ibid., 41–42.

15. Ibid., 44.

16. Ibid., 50–51.

17. Ibid., 50, n. 92.

18. Ibid., 59.

19. Ibid., 60, quoting Maximos the Confessor, *Scholia on the Divine Names* 4:33 (PG 4:305D).

20. Ibid., 61.

21. Ibid., 64.

22. Ibid., 90–91.

23. Ibid., 62, n. 126, makes this clear. In using later Byzantine perspectives to illuminate the Eastern patristic tradition in this way, Nellas perhaps suffers from a common tendency among Eastern commentators: that of assuming unanimity among writers of different periods that does not stand up to close scrutiny.

24. Ibid., 62.

25. Ibid., 63, n. 128, drawing attention to Chrysostom's *On Greeting Priscilla* 3:5 (PG 51:194).

26. Philip Sherrard, *Christianity and Eros: Essays on the Theme of Sexual Love* (London: SPCK, 1976), 26 (my italics).

27. In the Origenist tradition, as manifested in the thinking of Anthony the Great, for example, there was a stress on *to noeron*, the soul's intellectual state or intellectual nature. Rational beings were seen as having been created as a unique "intellectual substance" that was later broken up by the Fall into individualized portions, each endowed with a "heavy body." While aspects of Origenism were later seen as heretical, this concept often remained in Orthodox thinking. For Athanasius, for example, "Virtue consists of the soul preserving its intellectuality [*to noeron*] according to nature." For a discussion and references, see Norman Russell, "Bishops and Charismatics in Early Christian Egypt," in *Abba: The Tradition of Orthodoxy in the West—Festschrift for Bishop Kallistos (Ware) of Diokleia*, ed. John Behr, Andrew Louth, and Dimitri Conomos, 99ff (Crestwood, N.Y.: St. Vladimir's Seminary Press, 2003).

28. Nellas, *Deification in Christ*, 35.

29. Olivier Clément, *The Roots of Christian Mysticism: Text and Commentary*, 5th ed. (London: New City, 1998), 132.

30. See, for example, ibid., 134ff; cf. Jamie Moore, "Orthodoxy and Modern Depth Psychology," in *Living Orthodoxy in the Modern World*, ed. Andrew Walker and Costa Carras, 131ff (London: SPCK, 1996).

31. Vladimir Lossky, *The Mystical Theology of the Eastern Church* (Cambridge: James Clarke, 1957), 101.

32. Alexander Schmemann, *The Eucharist: Sacrament of the Kingdom* (Crestwood, N.Y.: St. Vladimir's Seminary Press, 1987), 33–34.

33. Philip Sherrard, "The Sacrament," in *The Orthodox Ethos: Essays in Honour of the Centenary of the Greek Orthodox Archdiocese of North and South America*, vol. 1, ed. A. J. Philippou (Oxford: Holywell Press, 1964), 134.

34. Ibid., 135.

35. Ibid., 139.

36. Ibid., 133–34.

Chapter 13

1. Stephen W. Need, "Rereading the Prologue: Incarnation and Creation in John 1.1–18," *Theology* 106 (2003), 403.

2. Ibid., 400.

3. Ibid.

4. My argument does not depend on the assumption that the *Logos* concept was central to the Fourth Gospel's proclamation. It may be true, as some have claimed, that the prologue is simply an existing hymn (or an adaptation of one) already used in part of the community that the gospel's author wished to influence. If this was so, it does not affect my argument. The important point is simply that the author saw this prologue as being at least consonant with what he wished to convey, and it was later taken up by the Christian community as a whole. In this sense, the exact "original intention" of the gospel's author is not important.

5. See Eugene TeSelle, "Divine Action: The Doctrinal Tradition," in *Divine Action: Studies Inspired by the Philosophical Theology of Austin Farrer*, ed. B. Hebblethwaite and E. Henderson, 79ff (Edinburgh: T&T Clark, 1990).

6. Ibid.

7. The panentheistic element of Maximos's view is emphasized in two essays in Philip Clayton and Arthur Peacocke, eds., *In Whom We Live and Move and Have Our Being: Panentheistic Reflections on God's Presence in a Scientific World* (Grand Rapids: Eerdmans, 2004). These essays are Kallistos Ware, "God Immanent yet Transcendent: The Divine Energies according to Saint Gregory Palamas," 157ff; and Andrew Louth, "The Cosmic Vision of Saint Maximos the Confessor," 184ff.

8. Louth, "The Cosmic Vision," 188.

9. Ibid.

10. Philip Sherrard, *Christianity: Lineaments of a Sacred Tradition* (Edinburgh: T&T Clark, 1998), 58.

11. Lars Thunberg, *Man and the Cosmos: The Vision of St. Maximus the Confessor* (Crestwood, N.Y.: St. Vladimir's Seminary Press, 1985), 75.

12. Need, "Rereading the Prologue," 403.

13. The different ways of relating incarnation and atonement have been admirably summed up in a classic twentieth-century text, G. Aulen, *Christus Victor: An Historical Study of the Three Main Types of the Idea of the Atonement* (London: SPCK, 1931).

14. Timothy Ware, *The Orthodox Church*, rev. ed. (Harmondsworth, England: Penguin, 1964), 203. We should note here, as Ware does, that this concept of the incarnation is not unknown in the West. For example, it is found in the works of Duns Scotus. It has never, however, entered the mainstream of thinking in the West as it has in the East.

15. Philip Sherrard, *The Rape of Man and Nature* (Ipswich, England: Golgonooza, 1993), 47.

16. John P. Dourley, *The Illness That We Are: A Jungian Critique of Christianity* (Toronto: Inner City Books, 1984), notes in particular that "Tillich could have counted on Jung's support in the view that the development of Christian theology since the introduction of Aristotle through Aquinas has been one leading consistently to the loss of the inner sense of God" (p. 34).

17. Sherrard, *The Rape of Man and Nature*, 49.

18. Ibid., 51–52.

19. Ibid., 106. Sherrard's account, although of great interest in what it affirms, is marred by being set within an ill-informed diatribe against modern science.

20. Philip Sherrard, *Christianity: Lineaments of a Sacred Tradition* (Edinburgh: T&T Clark, 1998), 58.

21. Ibid.

22. Ibid.

23. Need, "Rereading the Prologue," 403.

24. Some of the problems that are sometimes perceived are due to a one-sided interpretation of what Peacocke says. The best brief account of his approach may be that of D. R. Copestake, "Emergent Evolution and the Incarnation of Jesus Christ," *Modern Believing* 36:4 (1995), 27ff.

25. John Polkinghorne, *Scientists as Theologians: A Comparison of the Writings of Ian Barbour, Arthur Peacocke and John Polkinghorne* (London: SPCK, 1996), 70. Note that Peacocke himself does not accept this reading of his position.

26. Vladimir Lossky, *The Mystical Theology of the Eastern Church* (Cambridge: James Clarke, 1957), 101.

27. Alexander Schmemann, *The Eucharist: Sacrament of the Kingdom* (Crestwood, N.Y.: St. Vladimir's Seminary Press, 1987), 33–34.

28. Arthur Peacocke, *God and the New Biology* (London: Dent & Sons, 1986), 124.

29. Louth, "The Cosmic Vision," 189.

30. Ibid.

31. Ibid., 195.

Chapter 14

1. It is noteworthy in this respect that, among the mainstream churches, it has primarily been the Eastern Orthodox Church, with its theological and spiritual roots in the tradition I have described, that has taken major initiatives in this area. See, for example, the essays and statements collected in Alexander Belopopsky and Dimitri Oikonomou, eds., *Orthodoxy and Ecology: Resource Book* (Syndesmos, World Fellowship of Orthodox Youth, 1996).

2. Sallie McFague, *The Body of God: An Ecological Perspective* (London: SCM, 1993).

3. Ibid., 32.

4. Ibid.

5. Ibid., 35.

6. Olivier Clément, *The Roots of Christian Mysticism: Text and Commentary*, 5th ed. (London: New City, 1998), 132.

7. Panayiotis Nellas, *Deification in Christ: Orthodox Perspectives on the Nature of the Human Person*, trans. Norman Russell (Crestwood, N.Y.: St. Vladimir's Seminary Press, 1997), 41.

8. Ibid., 70.

9. Ibid., 71.

10. I have not been able to trace the origin of this phrase, but I believe it occurs in the writing of Kenneth Leach.

11. Ibid., 76.

12. Ibid., 149.

13. See especially ibid., 179ff.

14. Ibid., 182ff.

15. Mark I. Wallace, *Finding God in the Singing River: Christianity, Spirit, Nature* (Minneapolis: Fortress Press, 2005), 4.

16. Carl E. Braaten, "The Gospel for a Neopagan Culture," in *Either/Or: The Gospel or Neopaganism*, ed. Carl E. Braaten and Robert W. Jensen (Grand Rapids: Eerdmans, 1995), 7.

17. Wallace, *Finding God*, 17

18. Braaten, "The Gospel for a Neopagan Culture," 15.

19. Louis Bouyer, *Rite and Man: The Sense of the Sacral and Christian Liturgy* (London: Burns & Oates, 1963), 3.

20. Wallace, *Finding God*, preface.

21. Ibid., 10.

22. Celia E. Deane-Drummond, *Creation through Wisdom: Theology and the New Biology* (Edinburgh: T&T Clark, 2000).

23. Ibid., 89–90.

24. Ibid., 90–91.

25. Ibid., 115.

26. Ibid., 140.

27. As the translators of the *Philokalia* note in their glossary (under the term "intelligent"), the term *logikos* "is so closely connected with Logos . . . and therefore with the divine Intellect, that to render it simply as 'logical' and hence descriptive of the reason . . . is clearly inadequate." And as they note under the term "intellect" (*nous*) as it applies to humans, the notion of intellect must be carefully distinguished from the reason (*dianioia*), since "The intellect does not function by formulating abstract concepts and then arguing on this basis to a conclusion reached through deductive reasoning, but it understands divine truth by means of immediate experience, intuition, or 'simple cognition' (the term used by St Isaac the Syrian). The intellect dwells in the 'depths of the soul'; it constitutes the innermost aspect of the heart." See the glossary in *The Philokalia: The Complete Text Compiled by St. Nikodimos of the Holy Mountain and St. Makarios of Corinth*, vol. 4, trans. and ed. G. E. H. Palmer, Philip Sherrard, and Kallistos Ware (London: Faber & Faber, 1995), 432.

28. Gregory Palamas, "Topics on Natural and Theological Science," in *The Philokalia*, vol. 4, 361–62.

29. Deane-Drummond, *Creation through Wisdom*, 151.

30. The term "ones" here means "unites." The translation—which reflects the middle English of the original—is that in Julian of Norwich, *Showing of Love* (London: Darton, Longman & Todd, 2003), trans. Julia Bolton Holloway. The translator comments on this passage in her preface, p. xvii.

Chapter 15

1. Lars Thunberg, *Man and the Cosmos: The Vision of St. Maximus the Confessor* (Crestwood, N.Y.: St. Vladimir's Seminary Press, 1985), 75.

2. Stephen W. Need, "Re-reading the Prologue: Incarnation and Creation in John 1:1–18," *Theology* 106 (2003), 403.

3. Vladimir Lossky, *The Mystical Theology of the Eastern Church* (Cambridge: James Clarke, 1957), 101.

4. Because Byzantine authors were rarely systematic, it is in fact difficult to say whether they posited a unified approach to divine action in which a single model was sufficient. Although they did not speak of special divine action in the technical sense, they often spoke rather loosely in a way that might now be interpreted within that framework. In what follows, however, I advocate a neo-Byzantine model that firmly avoids the concept of special action. In this sense, it is inspired by the earlier Byzantine model but is not simply a restatement of it.

5. Kallistos Ware, "God Immanent yet Transcendent: The Divine Energies according to St. Gregory Palamas," in *In Whom We Live and Move and Have Our Being: Panentheistic Reflections on God's Presence in a Scientific World*, ed. Philip Clayton and Arthur Peacocke (Grand Rapids: Eerdmans, 2004), 160.

6. Andrew Louth, "The Cosmic Vision of St. Maximos the Confessor," in *In Whom We Live*, ed. Clayton and Peacocke, 189.

7. Willem Drees, "Thick Naturalism: Comments on Zygon 2000," *Zygon: Journal of Religion and Science* 35 (2000), 851.

8. This means that two kinds of complexity may be defined. The first relates to the way in which, for example, evolutionary biologists do not expect the past transformation of one species into another to be demonstrable in a repeatable way. They recognize that the complexity of the ecological niche that made this transformation possible could never be replicated. Here, the issue of repeatability is seen simply in terms of the practicability of reproducing the multifarious factors involved. The second, which is more conceptually subtle, relates to holistic factors of the sort that are often seen as being involved in the emergence of new "levels of complexity" in the cosmos, such as life and intelligent self-consciousness (which are discussed later in this chapter in relation to the insights of Arthur Peacocke). Here, the issue is tied to the sort of antireductionist approach—advocated, for example in Paul Davies, *The Cosmic Blueprint* (London: Unwin Hyam, 1987)—in terms of new "laws" or "organizing principles" that come into operation at each emergent level in nature's hierarchy of organization and complexity. It is not necessary, Davies argues, "to suppose that these higher level organizing principles carry out their marshalling of the system's constituents by deploying mysterious new forces for the purpose, which would . . . be tantamount to vitalism. . . . [Instead, they] could be said to harness the existing interparticle forces, rather than supplement them, and in so doing alter the collective behaviour in a holistic fashion. Such organizing principles need therefore in no way contradict the underlying laws of physics as they apply to the constituent parts of the system" (p. 143).

9. This point about extremity is underlined, as we have seen, by the existence of "regime change" phenomena such as superconductivity, which are scientifically demonstrable and understandable but represent discontinuities with "ordinary" experience that may neither have been predicted nor even have seemed possible before their demonstration. John Polkinghorne, *One World: The Interaction of Science and Theology* (London: SPCK, 1986), 74ff, has suggested this analogy as one that illuminates the character of events seen as miraculous.

10. See the essays in Clayton and Peacocke, eds., *In Whom We Live and Move.*

11. For an account of their advocacy of the causal-joint scheme, see, for example, Philip Clayton, *God and Contemporary Science* (Edinburgh: Edinburgh University Press, 1997), ch. 6. For their attitude to panentheism, see Clayton and Peacocke, eds., *In Whom We Live and Move.*

12. Drees, "Thick Naturalism," 851.

13. See the essays by Andrew Louth and Kallistos Ware in Clayton and Peacocke, eds., *In Whom We Live and Move.*

14. See, for example, John D. Barrow and Frank J. Tipler, *The Anthropic Cosmological Principle* (Oxford: Clarendon, 1986), 148ff.

15. To give just two examples, the determinism of Newtonian physics has disappeared entirely through the insights of quantum mechanics, and the ontology of the world is no longer understood in terms of a naively realistic understanding of the entities posited by scientific theory.

16. Although new arguments have arisen in relation to it, the most comprehensive account of this principle is still that of Barrow and Tipler, *The Anthropic Cosmological Principle.*

17. This general view does not, it should be noted, depend on an entirely naturalistic view of divine action. As we have seen in chapter 3, Arthur Peacocke continues to defend "special providence" in relation to aspects of divine action but sees the evolutionary emergence of humanity entirely in naturalistic terms. John Polkinghorne, too, makes a similar distinction, though with different boundaries.

18. See below for a discussion of this emergence in the thought of Arthur Peacocke.

19 Simon Conway Morris, *Life's Solution: Inevitable Humans in a Lonely Universe* (Cambridge, Cambridge University Press, 2003), 309–10.

20. Divine providence may be seen as "lawlike," I would argue, in the sense that identical situations will give rise to identical providential results (not because God is "constrained," but because God is consistent and reliable). For the reasons discussed above, however, such providence will rarely be predictable, since as we move to higher levels of complexity, situations become more difficult to identify reliably or to replicate. As we have seen in chapter 5, if we were to pursue this argument in patristic terms, then Augustine's implicit concept of a "higher" and a "lower" nature provides a good starting point.

21. For a view in terms of holistic organizing principles, see, for example, the comments of Paul Davies quoted in note 8 of this chapter. For more philosophically aware accounts, see Philip Clayton, *Mind and Emergence: From Quantum to Consciousness* (Oxford: Oxford University Press, 2004); and Dennis Bielfeldt, "The Peril and Promise of Supervenience for the Science-Theology Discussion," in *The Human Person in Science and Theology*, ed. Niels Henrik Gregersen, Willem B. Drees, and Ulf Gorman, 117ff (Edinburgh: T&T Clark, 2000).

22. Arthur Peacocke, *God and the New Biology* (London: J. M. Dent & Sons, 1986), 30.

23. For a fuller exposition of this and a criticism of aspects of Peacocke's own use of his insights, see Christopher C. Knight, *Wrestling with the Divine: Religion, Science, and Revelation* (Minneapolis: Fortress Press, 2001), ch. 4.

24 Morris, *Life's Solution,* 309.

25. As we have seen, Rahner's approach may be used to explore the nature of Christ's resurrection appearances. A good general introduction to this aspect of Rahner's thinking is Christopher F. Schiavone, *Rationality and Revelation in Rahner* (New York: Peter Lang, 1994).

26. For an exploration of this aspect of the character of theological language, see Knight, *Wrestling with the Divine*, chs. 7–8.

27. Philip Sherrard, *Christianity: Lineaments of a Sacred Tradition* (Edinburgh: T&T Clarke, 1998), 61–62.

28. Nancey Murphy, *Theology in an Age of Scientific Reasoning* (Ithaca, N.Y.: Cornell

University Press, 1990), has explained this Lakatosian term and indicated how it may be applied to theology.

29. Jacques Monod, *Chance and Necessity* (London: Collins, 1972), argues that the role of chance in the universe's evolution requires an atheistic understanding. In a teleological-christological approach, however, it is precisely the interplay of chance and necessity that allows the potential of the universe to be brought to fruition.

Chapter 16

1. See, for example, F. C. Happold, *Mysticism: A Study and an Anthology* (Harmondsworth, England: Penguin, 1963), 48: "The experiences of the mystics are not understandable unless one is prepared to accept that there may be an entirely different dimension from that of clock time or indeed of any other sort of time. For the mystic feels himself to be in a dimension where time is not, where 'all is always now.'"

2. Thomas Aquinas, *Commentary on Aristotle's Peri Hermeneias* 1:14, quoted in Brian Davies, *An Introduction to the Philosophy of Religion*, 2nd ed. (Oxford: Oxford University Press, 1993), 153.

3. For a more nuanced account of this statement, see Christopher C. Knight, *Wrestling with the Divine: Religion, Science, and Revelation* (Minneapolis: Fortress Press, 2001), ch. 5.

4. See, for example, C. J. Isham and J. C. Polkinghorne, "The Debate over the Block Universe," in *Quantum Cosmology and the Laws of Nature: Scientific Perspectives on Divine Action*, ed. Robert John Russell, Nancey Murphy, and C. J. Isham, 135ff (Vatican City: Vatican Observatory, 1993).

5. B. Davies, *An Introduction to the Philosophy of Religion*, 2nd ed. (Oxford: Oxford University Press, 1993), 141. The chapter from which this quotation comes provides an accessible account of the classical view and some of the reasons why it should not be abandoned.

6. Antje Jackelen, *Time and Eternity: The Question of Time in Church, Science, and Theology* (Philadelphia: Templeton Foundation, 2005), esp. ch. 2.

7. Vladimir Lossky, *The Mystical Theology of the Eastern Church* (Cambridge: James Clarke, 1957), 42.

8. "The Creed of St. Athanasius," in the 1662 version of the Church of England's *Book of Common Prayer* (punctuation modernized).

9. Ibid.

10. Lossky, op.cit., 43.

11. Ibid.

12. Ibid., 45–46.

13. J. R. Lucas, "The Temporality of God," in *Quantum Cosmology and the Laws of Nature*, ed. Russell, Murphy, and Isham, 236.

14. John Polkinghorne, *Science and Christian Belief: Theological Reflections of a Bottom-Up Thinker* (London: SPCK, 1994), 78–79.

15. Philip Clayton, *God and Contemporary Science* (Edinburgh: Edinburgh University Press, 1997), 220–27.

16. A. R. Peacocke, "God's Interaction with the World: The Implications of Deterministic 'Chaos' and of Interconnected and Interdependent Complexity," in *Chaos and Complexity: Scientific Perspectives on Divine Action*, ed. R. J. Russell, N. Murphy, and A. R. Peacocke (Vatican City: Vatican Observatory, 1995), 283.

17. John Polkinghorne, *Scientists as Theologians: A Comparison of the Writings of Ian Barbour, Arthur Peacocke, and John Polkinghorne* (London: SPCK, 1996), 26ff.

18. See, for example, Jürgen Moltmann, *The Crucified God* (London: SCM, 1974); W. H. Vanstone, *Love's Endeavour, Love's Expense* (London: Darton, Longman & Todd, 1977); and A. R. Peacocke, *Theology for a Scientific Age: Being and Becoming—Natural, Human and Divine*, rev. ed. (London: SCM, 1993), 123ff. For a comparable influence on another British theologian, see Derek Stanesby, *Science, Religion and Reason* (London: Routledge, 1985), 128ff.

19. Willem Drees, "Thick Naturalism: Comments on *Zygon* 2000," *Zygon: Journal of Religion and Science* 35 (2000), 851.

20. For a brief account of the neo-Thomist approach, see Ian G. Barbour, *Issues in Science and Religion* (London: SCM, 1966), 425ff.

21. John Polkinghorne, *Faith, Science and Understanding* (London: SPCK, 2000), 116.

22. Bad statistics are evident, for example, in the claim that in a random gathering of only thirty people, it would be extraordinary if two of them shared the same birthday. To the nonmathematician, this coincidence may seem unlikely, but in fact, in a gathering of this size, the probability of it occurring is greater than 50 percent.

23. Barbour, *Issues in Science and Religion*, 428.

24. Russel Stannard, "God in and beyond Space and Time"; in Philip Clayton and Arthur Peacocke, eds., *In Whom We Live and Move and Have Our Being: Panentheistic Reflections on God's Presence in a Scientific World* (Grand Rapids: Eerdmans, 2004), 116.

Afterword

1. See for example, Philip Sherrard, *The Rape of Man and Nature* (Ipswich, England: Golgonooza, 1993).

2. If modern science constitutes the chief of Sherrard's *bêtes noires*, then not far behind it comes the Western Christian tradition in its various forms. His comments on this tradition are often illuminating, but they also often manifest his inability to see good where good exists.

3. Philip Sherrard, *Christianity: Lineaments of a Sacred Tradition* (Edinburgh: T&T Clarke, 1996), 61–62.

4. Ibid., 63.

5. Ibid.

6. Ibid.

7. John Behr, "Faithfulness and Creativity," in *Abba: The Tradition of Orthodoxy in the West—Festschrift for Bishop Kallistos (Ware) of Diokleia*, ed. John Behr, Andrew Louth, and Dimitri Conomos (Crestwood, N.Y.: St. Vladimir's Seminary Press, 2003), 174.

8. Timothy Ware, *The Orthodox Church*, rev. ed. (Harmondsworth, England: Penguin, 1997), 198.

Index